Emilian Mihailov, Tenzin Wangmo, Victoria Federiuc, Bernice Elger (Eds.)
Contemporary Debates In Bioethics: European Perspectives

Emilian Mihailov, Tenzin Wangmo,
Victoria Federiuc, Bernice Elger (Eds.)

Contemporary Debates in Bioethics

European Perspectives

sciendo

ISBN 978-3-11-057120-2
e-ISBN 978-3-11-057121-9

Published by Sciendo (De Gruyter's brand for publishing services)
Subsidiary of Walter de Gruyter GmbH, Berlin
The book is published with open access at www.degruyter.com.

Library of Congress Cataloging-in-Publication Data
A CIP catalog record for this book has been applied for at the Library of Congress.

www.degruyter.com
Cover illustration: berya113

Contents

Foreword —— IX

Christopher Gyngell

1 Good Parents and New Reproductive Technologies —— 1

1.1 Introduction —— **1**

1.2 Use of RGTs —— **2**

1.2.1 IVF and PGD as tools —— **2**

1.2.2 Gamete screening —— **3**

1.2.3 Genetic engineering —— **3**

1.2.4 *In vitro* gametogenesis —— **5**

1.3 Should parents influence the genes of their children? —— **5**

1.4 Does the method matter? —— **7**

1.5 Conclusion —— **8**

References —— **9**

Bogdan Olaru

2 Bypassing Morality Through Conventional and Unconventional Forms of Moral Enhancement —— 11

2.1 The argument emanating from similarity —— **11**

2.2 Selecting for particular ways of reasoning, disruptions of one's narrative identity, and freedom from any mental manipulations —— **14**

2.3 Fundamental similarity between conventional and nonconventional moral enhancement —— **16**

2.4 The constraint of moral equality between the enhancer and the enhanced person —— **18**

References —— **22**

Marcello Ienca

3 The Neuroenhancement Continuum and the Minimal Rule —— 23

3.1 Introduction —— **23**

3.2 Nootropics and the origin of human neuroenhancement —— **24**

3.3 The neuroenhancement continuum: nootropics and other enhancers —— **26**

3.4 The neuroenhancement continuum II: cognition and other systems —— **27**

3.5 A minimal rule for the administration of neuroenhancers: MiRNA —— **29**

3.6 Application of the MiRNA —— **30**

3.7 Safety, self-determination, equality, and information —— **32**

3.8 Conclusion — **33**
References — **35**

Blanca Rodríguez López

4 Procreative Beneficence: is Selection Really Better Than Genetic Modification? — 37
4.1 Introduction — **37**
4.2 Some considerations on the PPB — **38**
4.2.1 Definition of enhancement — **38**
4.2.2 The best life — **38**
4.3 Selection — **39**
4.3.1 Two (unconvincing) explicit reasons — **40**
4.3.2 One (unconvincing) implicit reason? — **41**
4.3.3 The real (explicit) reason — **42**
4.4 PPB and the nonidentity problem — **43**
4.5 Why selection is not morally superior — **44**
4.6 Conclusion: a defense for an extension of PPB — **46**
References — **47**

Daniela Rusnac

5 Practical Ethics Issues in Gene Therapy and Genetic Testing — 49
5.1 Introduction — **49**
5.2 Genetic testing — **49**
5.3 Gene therapy — **53**
5.3.1 Basic and preclinical research — **55**
5.3.2 Clinical use: somatic — **55**
5.3.3 Clinical use: germ line — **55**
5.4 Conclusion — **56**
References — **56**

Emilian Mihailov

6 Refocusing the Nudge Debate on Organ Donation — 58
6.1 Introduction — **58**
6.2 The success of opt-out systems — **59**
6.3 Tempering the opt-out enthusiasm — **61**
6.4 Opt-out policy and consent standards — **64**
6.5 Beyond the dichotomy of presumed and explicit consent — **66**
6.6 Nudging beyond potential donors — **68**
6.7 Conclusion — **69**
References — **69**

Mihaela Frunză, Iulia Grad, Sandu Frunză, Ovidiu Grad

7 CLICK HERE! To Find More About Organ Transplantation: Ethical Aspects of Media Stories on Organ Donation from Romanian Newspapers — 72
7.1 Introduction — **72**
7.2 Sensational stories on organ donation in the Romanian paper-based media (2008–2012) — **74**
7.3 "Look, mom, now I've got hands!" Unlikely stories of transplantation from recent Romanian periodicals — **75**
7.4 From mutual ignorance to mutual partnerships: students' campaigns promoting organ donation — **81**
7.5 Conclusions — **81**
References — **82**

Mihaela Constantinescu

8 Seeing the Forest Beyond the Trees: A Holistic Approach to Health-Care Organizational Ethics — 86
8.1 Introduction — **86**
8.2 Organizational ethics in health care — **87**
8.3 Compliance versus integrity: an apparent dichotomy — **88**
8.4 Embedding ethics in HCOs: optimal alignment between culture and structure — **91**
8.5 Conclusion and suggestions for future research — **92**
References — **94**

Eva De Clercq

9 Disability @ the Movies: Toward a Disability-Conscious Bioethics — 97
9.1 Introduction — **97**
9.2 What *can* a body *do*? — **99**
9.3 Moral imagination at the movies — **100**
9.4 Disability stereotypes on the silver screen — **102**
9.5 Disability in contemporary cinema: from isolation to inclusion? — **103**
9.6 Conclusion: the power of re-signification — **105**
References — **106**

Constantin Vică

10 The Info-Computational Turn in Bioethics — 108
10.1 All watched over by machines of loving grace? — **108**
10.2 The digital epidemiology — **110**
10.3 Ways of understanding the info-computational turn — **113**
10.4 Conclusions — **116**
References — **118**

Adriana Paladi, Victoria Federiuc

11 The Principle of Autonomy in Palliative Care: the Moldavian Perspective — 121

11.1 Introduction — **121**

11.2 Principle of autonomy — **122**

11.3 Moral requirement of patient information — **123**

11.4 Some reasons for covering up bad news — **124**

11.5 The case of the Republic of Moldova — **125**

11.6 Delivering bad news in a sensitive manner — **128**

11.7 Conclusion — **129**

References — **130**

Laurenţiu Staicu, Octavian Buda

12 Philosophical Foundations and The Role Of Counseling in The Ethics of Informed Consent — 132

12.1 Introduction — **132**

12.2 Informed consent as an expression of liberty — **133**

12.3 Informed consent as an expression of morality — **136**

12.4 A moral dilemma and its philosophical presuppositions — **139**

12.5 Conclusions — **143**

References — **144**

Cristian Iftode

13 Bioethics as Biopolitics: A Foucauldian Perspective — 145

13.1 Introduction — **145**

13.2 Madness, the asylum institution, and the psychiatric power — **146**

13.3 Biopolitics as “power’s hold over life”: the three levels of an analysis of “normalization” — **148**

13.4 In search of a new “ethics of life”: ethical subjectivation vs. political subjection — **154**

13.5 Final remarks — **157**

References — **158**

Isabelle Wienand, Milenko Rakic, Sophie Haesen, Bernice Elger

14 How Should One Die? Nietzsche’s Contribution to the Issue of Suicide in Medical Ethics — 160

14.1 Introduction — **160**

14.2 The hermeneutic challenge of HTH §185 — **162**

14.3 The religious interpretation of suicide — **165**

14.4 Conclusion — **167**

References — **167**

Contributors — 169

Foreword

Bioethical problems are almost never settled completely due to the varying contexts and cultures under which they play out. Yet, progress is possible. This does not mean that appropriate solutions are not available for each context and culture, nor does it mean that they are imperfect. Bioethical issues have a special dynamic, which makes them open ended in the context of the diversity and the changing human experiences. Competing moral values can be harmonized in some contexts and cultures, but hardly in others. Scientific advances and breakthroughs open new possibilities for action, which equally liberates and burdens human decision-making. And even when there is an overlapping consensus with regard to the adequacy of normative frameworks, there will still be thorny questions about their application, for which local and general empirical knowledge needs to be generated.

Moral disagreements and uncertainties keep bioethics open to sharpening and deepening of the normative and empirical investigations. Philosophical analysis of arguments and presuppositions will have to determine to which extent standard concepts and principles are transferable to new situations and other contexts, evaluate critically the ethical implications of biomedical policies that prove successful, or mark out irrational attitudes against scientific advances with significant potential in terms of social benefits. Empirical research will have to identity in what conditions the application of morally desirable policies can backfire, and, consequently, determine more favorable conditions or document people's evaluative attitudes and explain by what they are driven.

Contemporary Debates in Bioethics: European Perspectives aims to contribute to these challenges. It includes 14 chapters by philosophers and social scientists on issues of enduring and contemporary importance, such as organ donation, biomedical enhancement, genetic editing, euthanasia, informed consent, biopolitics, social interpretation of disability, end of life, health-care organizational ethics, and the convergence of computing, information, and biological life. The articles feature rigorous analysis and the original results that will be of interest to scholars working on these important issues. Furthermore, they bring important discussions to the table and are topics that need in-depth analysis.

Enhancement technologies constitute one of the most controversial topics in contemporary bioethics. There seems to be a widespread belief that interventions to treat diseases are significantly more permissible than interventions aimed at improving normal capacities, and moreover, that traditional means of improvement (education and institutions) are less problematic than pharmacological, genetic, or biotechnological interventions. These moral asymmetries are pervasive in the public debate and popular culture, but they face philosophical problems. The debate around enhancement technologies deserves an advanced analysis in order to clear it away from gut feelings but also to avoid blindly accepting promising biotechnologies.

In the face of severe organ shortage, there are many proposals to improve organ availability, which needs to be evaluated from a normative standpoint. The proposed policies may be in tension with some ethical standards, but they could still promote other moral values. This gives rise to dilemmas in which one has to balance competing principles or adjust the policies in order to satisfy the relevant ethical standards. In addition, considerate public discourse about organ transplantation is essential in many cases to overcome the ethical dilemmas surrounding organ donation decisions. The way in which organ donation and transplantation are depicted in the media and public debates could widely influence people's attitudes and decisions, potentially undermining the hard work of scholars and policymakers.

The role of films in fruitfully engaging the public about the life experiences of persons with disability, as well as the use of this medium to launch a more-nuanced education and public debate, is an interesting challenge for bioethicists. The topic of disability and movies brings forth the interconnectedness between public opinions and individual circumstances, as well as its resulting choices. In a similar vein, the influence of macrolevel structures on microlevel individual decisions is also underscored in the chapter on health care organization ethics. Thus, for better and greater evidence-based medical decisions, physicians require an appropriate formal environment and culture that allows them to make unbiased choices that are in the best interest of the patient.

This volume also includes topics that may be of concern to the specific niche of bioethicists. However, these topics will gain relevance in the future in the context of globalization and technological innovations. For instance, we know very little about palliative care and the implementation of the autonomy principle in medical contexts in less-resourced countries. The literature to date mostly provides information from, for instance, the US, Western Europe, Australia, and Canada. Hence, a critical evaluation of autonomy in palliative care in the context of Moldova is a unique contribution. That technology is changing how we view and understand the world is becoming more and more evident with its pervasiveness in our day-to-day interactions, but whether bioethics is ready to redefine human agency in our time and era of data science brings us to unexplored territory.

Bioethics has struggled for a long time to harmonize the concept of "informed consent", with pragmatic limitations on how to obtain informed consent in the clinical or research context. There are shortcomings in the "thin" framework of respect for patient autonomy, whereby a doctor presents the patient with information and, possibly, a recommendation about what is medically indicated but then leaves the moral decision-making to the patient. On the other hand, "thick" frameworks prescribe a doctor's duty to build the patient's moral conscience, or the doctor and patient should both decide based on virtues such as prudence or temperance. However, if the doctor and patient do not come from the same (religious or not) moral communities, a dilemma results because they would not agree on values and the doctor would remain an information presenter who does not provide any moral support but leaves the decision-making entirely to the patient. Given such a deep disagreement, an interesting question arises as to whether philosophical counseling could explain and make explicit the metaphysical worldviews involved, and, consequently, mediate the process of obtaining informed consent.

Nietzsche died on August 25 in 1900. He lacked any opportunity to witness or imagine the possibilities of modern technologies, such as organ transplantation, intensive care, or sophisticated palliative pain and symptom treatments, including deep continuous sedation. This volume contains two chapters that offer interpretations of philosophers who died between decades and centuries ago. Why is it judged important to do so in order to enlighten the discussion of current bioethical topics? The need to refer to the thoughts of philosophers from different times and backgrounds results from the particular characteristics of modern bioethics thinking that started in the US in the 1970s. Indeed, as today's bioethics discourse takes place in a pluralistic society, there is a tendency to prefer "thin" bioethics and to refrain from thick concepts. One typical example of what is referred to as a "thin" framework is principlism. Bioethics in this type of framework is often reduced to procedural ethics or to a way to solve conflicts in society between the "thick" frameworks of each philosophical or theological moral community by simply giving the last word to patients whose autonomy must be respected. In this context, two chapters fill the gap of going back to thick concepts that have been developed by influential philosophers, within their own specific frameworks developed within historical culture and contexts.

Overall, the volume hopes to capture a European gist of theoretical sensibilities, conceptual resources, and research interests, but not in an adversarial way, as opposed to American bioethics. Indeed, bioethics has been dominated by American scholars. However, under the rapid globalization of bioethical issues, it is much harder to draw sharp distinctions between what is European bioethics and what is not. Instead, the volume gathers contributions from European scholars as they collaborate and form a research network, drawing on a diversity of philosophical traditions and local knowledge, with the aim of debating universal bioethical problems.

Emilian Mihailov
Tenzin Wangmo
Victoria Federiuc
Bernice Elger

Christopher Gyngell

1 Good Parents and New Reproductive Technologies

1.1 Introduction

Humans have always had the ability to influence the genetic makeup of their children. Individuals who wanted tall and attractive children, for instance, could find tall and attractive partners to reproduce with, thereby raising the probability that their progeny would be tall and attractive. However, until very recently, this power was limited. Individuals were often not lucky enough to have a wide range of sexual partners to choose from. Even if they did, there was no guarantee that their children would inherit the traits that were desired.

The past few decades have seen a rapid increase in the power of parents to influence the genetic makeup of their children. Since the 1990s, a range of biotechnology tools have been available, which give parents some degree of control over the genetic makeup of their children. Such technologies are rapidly expanding. In April 2015, it was announced that the gene editing (GE) technique, clustered regularly interspaced short palindromic repeats (CRISPR), had been used to make edits in human embryos for the first time. The study was conducted in China on the disease beta thalassemia – with mixed success (Liang et al. 2015). In February 2016, the UK became the first country to officially approve GE research in human embryos. The decision means experiments in which the genes of embryos are manipulated will likely begin in the UK in 2017 (Callaway 2016).

The development of GE has been marred by controversy. Some public interest groups, including the United Nations Educational Scientific and Cultural Organization (UNESCO), have called for an international ban on any GE research in human embryos. The US-based National Institutes of Health maintained that performing such research would cross "a line that should not be crossed" (Collins 2015, 1). The major scientific journals *Nature* and *Science* have published commentaries, which call for this research to be strongly discouraged or stopped altogether (Lanphier et al. 2015; Baltimore et al. 2015).

While GE is controversial, other techniques that allow parents to influence the genetic makeup of their children are relatively common. Sperm banks and egg donation websites allow women and couples to pick and choose between different gamete donors based on a range of characters – and then create a child through in vitro fertilization (IVF) or artificial insemination. In the US alone, between 30,000 and 60,000 children are born through sperm donation each year and >8,000 from egg donation (Sabatello 2015). Preimplantation genetic diagnosis (PGD) allows embryos created through IVF to be tested for the presence or absence of genetic conditions, before implantation. It is currently possible to use PGD to select against any one of nearly 400 conditions (Human Fertilisation and Embryology Authority [HFEA] 2017).

In this paper, I first introduce the range of "reproductive genetic technologies" (RGTs) – technologies that allow parents to influence the genetic makeup of their future children.[1] I then examine two ethical questions: (1) Assuming such technologies are safe and effective, do parents have reasons to influence the genes of their children? (2) If so, does the type of RGT they use matter, legally and morally?

1.2 Use of RGTs[2]

1.2.1 IVF and PGD as tools

The development of IVF in the 1970s marked an important milestone in human reproduction. For the first time in history, human embryos could be created outside the body of the mother. This innovation was followed in the early 1990s by PGD, in which the embryos created in vitro were tested for the presence or absence of particular genes (Theodosiou and Johnson 2011).

PGD was initially developed as an alternative to prenatal testing and selective abortion within a legally and medically defined boundary (Theodosiou and Johnson 2011). It allowed parents to create many embryos and test each for genes associated with serious disabilities. This allowed parents to avoid serious disability without the emotional and physical costs associated with abortion. However, the potential for IVF and PGD to be used for nonmedical purposes soon became apparent. Through IVF, it is possible for parents to choose between embryos based on the presence of genes associated with nonmedical traits, such as height or intelligence. In theory, choosing an embryo that is likely to be taller, for instance, comes at little extra cost to the mother (although regulations in many parts of the world prevent such nonmedical applications).

Many human traits have a strong genetic component and could thus potentially be targeted though PGD. Geneticists have already identified genes associated with height (Berndt et al. 2013), certain personality types (Jang et al. 1996), intelligence (Desrivières et al. 2015), and musical ability (Oikkonen et al. 2015). As our knowledge of genetics increases, it will likely become possible to perform quite sophisticated genetic analyses on embryos before implantation. Hence, IVF and PGD potentially provide parents with a mechanism to influence a great variety of traits in their children.

1 This is independent of whether the identities of their children are also affected.

2 This section draws on my previous work: C. Gyngell and M. Selgelid. 21st Century Eugenics, In *The Oxford Handbook of Reproductive Ethics*, Leslie Francis (ed.) Oxford University Press (2016): 141–158.

1.2.2 Gamete screening

Gamete screening (GS) involves selecting between gamete donors, based on heritable characteristics. For decades, in countries such as the USA, individuals have been able to influence the genes of their children by using donated gametes (sperm and eggs) in combination with technologies such as IVF. Those using donated gametes often have access to detailed medical histories of the donors. Such individuals can not only reduce the chances of their children having a genetic disease but also choose donors with good medical histories. GS can not only be used to select against disabilities, it can also be used to select for nondisease traits such as eye color and height. Companies such as *Elite Egg Donors* allow prospective parents to choose between various gamete donors based on a wide variety of factors, including education, weight, and ethnicity.

While GS is currently relatively imprecise, new technologies that are likely to increase its power are being developed. In 2013, the genomics company '23andMe' received a patent for a technology called "Gamete Donor Selection Based on Genetic Calculations". While there are no suggestions that this technology is currently used, in the future, it would allow individuals accessing assisted reproductive services to choose between sperm or egg donors based on the statistical likelihood of the resulting child having a certain phenotype. Using 23andMe's technology, a woman wanting a blue-eyed child could select between different sperm donors to maximize this probability. She would differentially create an embryo with desirable characteristics, rather than selecting between different embryos or modifying an existing one.

1.2.3 Genetic engineering

In addition to being able to select between embryos based on their genetic makeup, parents may soon have the ability to directly modify embryos using genetic engineering technologies. Genetic engineering technologies potentially allow new genes to be inserted into embryonic DNA (Liang et al. 2015) and existing genes to be modified or deleted. These technologies could potentially be used to create much more significant changes in the traits of children than is possible through GS and PGD.

Genetic engineering technologies have been successfully used in other species to alter their physical, cognitive, and social characteristics. For example, in 2007, scientists at Case Western Reserve University used genetic engineering technologies to alter a gene called "PEPCK-A" in mice. The resulting transgenic mice could run for 6 km without a break – 30 times longer than a normal mouse's limit of 200 m. They also had extended life spans, compared to their unaltered counterparts, and retained the ability to breed well into old age (Hakimi et al. 2007). In 1999, scientists engineered mice to overexpress the gene "NR2B", which codes for a nerve cell receptor. This was shown to lead to dramatic improvements in memory, with transgenic mice being able to remember objects and experiences for many days longer than unaltered mice (Tang et al. 1999). The social characteristics of some animals have also been altered using genetic engineering technologies. Polygamous voles can be turned monogamous by modifying genes associated with the vasopressin V1a receptor (Lim et al. 2004).

Early techniques relied on viruses to deliver novel genetic material to cells. This method was too ineffective and imprecise to have serious potential as a clinically useful modifying tool of human DNA. It often only changed one of the two copies of the target gene, meaning animals had to be bred together to make modifications effective. This method also made unintended changes to large segments of the genome, and only a small proportion of the modified animals did not suffer serious side effects.

Recently, a revolution in genetic engineering began. In 2012, a laboratory led by Jennifer Doudna and Emmanuelle Charpentier showed that a molecule used as part of the bacterial immune system could be used to edit DNA (Jinek et al. 2012). The CRISPR–CRISPR-associated protein (Cas)9 system contains two parts, a molecule that binds to particular DNA sequences (CRISPR) and an enzyme that cuts the DNA (Cas9). In a very short space of time, the GE technique CRISPR has been used to make precise and heritable changes to animals. It has been used to create malaria-fighting mosquitoes, drought-resistant wheat, hornless cows, and monkeys with targeted mutations. The potential applications of GE in a decade are difficult to imagine. Because of their increased precision, GE techniques are the first genetic engineering technologies to have serious potential to modify human DNA.

It is clear that germline GE holds tremendous potential in the fight against many types of diseases. Most immediately, GE could be used to correct mutations that cause simple genetic diseases, such as cystic fibrosis, muscular dystrophy, and Tay–Sachs disease. Currently, such diseases can largely be prevented through genetic selection technologies such as PGD (HFEA 2017). However, PGD has significant limitations. Its ability to avoid disease is directly related to the number of embryos that can be created through IVF. Sometimes, couples will produce only one or two embryos, in which case PGD will not be effective in avoiding even simple genetic diseases. GE can be used to make multiple changes to a single embryo. It is free of PGD's limitations and would be a more efficient way of preventing simple genetic diseases.

More significantly, GE's ability to make multiple changes to a single embryo means that, in the long term, it could be used to prevent a far greater range of diseases than PGD. Cancer, diabetes, and heart disease all have significant genetic components. It is at least conceivable that we could use GE to make us resistant to these diseases – which are among the leading cause of mortality worldwide (World Health Organization [WHO] 2017). Imagine a steroid injection was developed, which if taken by a woman while pregnant would change the in-uterine environment in such a way that the embryo becomes resistant to cancer and cardiovascular disease. Most would consider such an intervention to be a wondrous medical breakthrough, which should be provided to all. GE may make such an intervention a reality one day.

In the future then, it seems plausible that it will be technologically possible for parents to modify the DNA of embryos created through IVF. As genetic technologies continue to advance, it may also become possible for embryos conceived naturally to be modified via vectors delivered directly to the uterus. This would not only provide a novel way for parents to be able to treat and prevent disease, it could also allow parents to influence a wide range of nonmedical characteristics of their children.

1.2.4 *In vitro* gametogenesis

Technologies that promise to greatly increase the selective power of IVF and PGD are currently being developed. In vitro gametogenesis (IVG) involves artificial production of germ cells (oocytes and sperms) from other cell types. It is now possible for embryonic murine and adult murine stem cells to be turned into sperm or egg cells in a Petri dish (Magnusdottir and Surani 2014). These cells are functional and have resulted in the birth of fertile offspring. This technology potentially allows any adult to generate thousands of germ cells from the stem cells contained in their bone marrow.

While most IVG research to date has been carried out in mice, there are several interesting possibilities for reproductive medicine in humans. The technique could one day allow people with no functional germ cells to produce children using standard IVF techniques. The technique could therefore have significant therapeutic benefits for those who are infertile.

In addition, the technique could greatly increase the selective power of IVF and PGD. The power of IVF is currently limited by the number of viable embryos a couple can produce (Bourne et al. 2012). Using IVG, any woman could potentially make hundreds or thousands of oocytes from her somatic cells. These could be used to make hundreds of embryos, all of which could undergo PGD. This would greatly increase the ability of IVF and PGD to be used to target polygenic traits – or multiple genetic traits at the same time. For example, imagine that 20 different genes contribute to a particular trait. If a couple aims to use PGD to select for 20 different alleles in an embryo, they would need to create around 10,000 embryos to make it sufficiently likely that one will have the desired combination at all 20 loci (Bourne et al. 2012). This is impossible through IVF and PGD today but could become possible in the future through IVG.

1.3 Should parents influence the genes of their children?

If these aforementioned technologies were safe, legal, and available to use on a developing embryo, would it be ethical to use them? I think it would be. I believe that we have strong moral reasons to use RGTs to protect future children from disease and suffering, such as those caused by single-gene disorders such as Tay–Sachs, spinal muscular atrophy, cystic fibrosis, and so on. Changing your child's genes to improve their health is consistent with good parenting.

It is commonly accepted that we should prevent disease in our children through nongenetic methods. For example, we expect pregnant women to eat well during pregnancy and avoid risky behaviors, if this will benefit the health of future children. We think this even when their actions will change the identity of the child who is born – by changing which egg is fertilized by which sperm. To explain, take the following case, adapted from Derek Parfit (1992, 358). Imagine a woman who wants to have a child, but who has an infectious disease. If she gets pregnant while she has the disease, her child will have a serious birth defect. If she waits 6 weeks, until her disease has passed, her child will be healthy. Most would agree that, as a parent, individuals in this position have good reason to wait the 6 weeks, even though it changes the identity of the future child.

So it is clear that parents try to benefit their future children through nonenvironmental measures. Why would genetic measures be morally different? Indeed, it is possible that some prenatal supplements currently widely taken do in fact have a genetic effect. For example, many women take vitamin D supplements through pregnancy. This is believed to make the developing child more resistant to bone diseases such as rickets (Royal Children's Hospital [RCH] 2017). We generally think pregnant women who take vitamin D supplements are acting responsibly. But vitamin D may alter the genes of the developing child – by changing the way its genes are expressed (Bocheva and Boyadjieva 2011). We think that women should take vitamin D supplements in pregnancy; therefore, we think it is permissible to make some genetic changes to a developing child in order to improve their health.

GE technologies make a different sort of genetic change to an embryo than vitamin D supplements. They change the genetic sequence, rather than the way genes are expressed. However, suppose that we discover that vitamin D supplements actually did work by changing the genetic sequence of a developing embryo? Would this discovery mean that we should advise women to stop taking vitamin D supplements during pregnancy? I think not. What is important is that taking vitamin D supplements while pregnant protects the developing child from developing diseases later in life. The mechanism through which this is achieved is irrelevant.

Some argue that the fact that RGTs make heritable change means that parents should not use those (Lanphier et al. 2015). But it is unclear why the fact that a disease is heritable should provide reasons against correcting it. To see why, imagine a hypothetical genetic disease that causes a hole to develop in a baby's heart soon after birth. The condition is nearly fatal. A new treatment (T1) involves injecting enzymes directly in the heart after birth to prevent the hole from forming. This treatment cures 80% of cases and has a low risk of side effects.

There are no reasons why T1 is unethical. Indeed, it is a moral imperative to provide T1.

Now imagine that babies who will get this condition can be detected by an in-utero genetic test. Treatment T2 injects the exact same enzymes as T1 but does so at the embryonic stage. This prevents the hole from forming in 80% of cases, and it has a low risk of side effects.

Are there any morally relevant differences between T1 and T2? The only difference between the treatments is that T2 is applied in utero and T1, soon after birth. This seems morally irrelevant. Imagine that we can only fund T1 or T2, and that T2 cures 85% of cases rather than 80% as in T1. In this case, it seems that we now have decisive reasons to prefer T2 over T1.

One might argue that the fact that T2 produces a heritable change, unlike T1, is morally relevant and counts against T2. I believe this reasoning is flawed. As T2 corrects the genes that cause this disease, individuals who have T2 will not be at risk of passing on the disease to their children. Individuals who get T1 remain at risk of having children with the same disease. This is an advantage of T2, not a cost.

It is possible to use genetic engineering technologies and other RGTs to make a small number of genetic changes to protect our children from disease. The total length of the human genome is >3 billion base pairs. In some cases, changing just one base pair in an embryo will prevent diseases later in life. This means that changing <0.0001% of your child's genome can protect him/her from a suffering. This seems clearly ethically justified.

Another objection to RGTs stems from the fact that they could be used not just to protect children from disease but that they also influence children's other traits. They could be used as a tool of enhancement, rather than just disease prevention. The ethics of enhancement are complex and have been discussed at length elsewhere (e.g., Buchanan and Brock 2007). I will take no stand on that issue in this chapter. But even if there is a sustained universal objection to human enhancement, this does not change the moral reasons we have for preventing disease. We could limit the use of RGTs to disease prevention, just as we currently do with PGD. Hence, there do seem strong moral reasons to use RGTs to change the genetic makeup of children in order to prevent disease.

1.4 Does the method matter?[3]

If the argument in the preceding section is accurate, parents have moral reasons to use RGTs in order to prevent disease. Does it matter, both legally and ethically, which RGT they use? Is the use of GS ethically the same as selection or GE?

If we just look at the legal status of these technologies, it looks as if there are significant differences. GS is widely unregulated. Companies such as Elite Egg Donors (discussed earlier) have clinics in diverse countries, such as Mexico, Barbados,

3 This section is adapted from a short article that I wrote for *The Conversation*. Refer C. Gyngell. If you can screen for brown eyes, you should be able to edit out genetic disease. *The Conversation* July 11, 2016. Available at https://theconversation.com/if-you-can-screen-for-brown-eyes-you-should-be-able-to-edit-out-genetic-disease-61927.

California, and Nepal (*Elite Egg Donors* website http://eliteeggdonors.com/ 2017). In 2015, the London Sperm Bank was criticized for its decision to ban sperm donors who suffer from minor neurological disorders, including dyslexia and Asperger syndrome. Not only can GS be used to select against mild disabilities such as dyslexia, it can also be used to select for nondisease traits such as eye color and height.

PGD is far more widely regulated than GS. In many parts of the world, PGD is limited to the prevention of disease. For example, in the UK, regulations limit PGD to being used to select against "serious" inherited conditions. However, what is regarded as "serious" is considered on a case-by-case basis. Each proposed use of PGD is examined individually. Those that are risky or frivolous can be rejected.

The reproductive use of GE is widely banned around the world (including in the UK, Australia, Canada, Mainland Europe, and Japan (Isasi et al. 2016) either through legislation or best practice guidelines. Such bans make sense considering that GE technologies are so new and are only now beginning to be used in animal and somatic cells. There may still be consequences of GE that we do not understand, and much more research into the safety of genome editing needs to be performed before it is considered for reproductive use. It is important to note that such laws do not just ban unsafe uses of GE technologies – but any use. GE is developing rapidly. At some point, GE technologies are likely to be widespread, cheap, safe, and precise. If GE becomes a safe technology, will it be ethically different from other RGTs?

It seems inconsistent to hold that the *method* we use to select for or against an inherited condition should make such a significant difference to its legal status. A woman wanting to have a green-eyed child with blond hair, low predisposition to obesity, and high intelligence is legally able to use GS to choose among potentially thousands of donors, based on which of the donors is most likely to give her the child she desires. Conversely, a couple who wants to use GE technology to save their child from developing Tay–Sachs disease – a degenerative disease that results in death by the age of 3 years, cannot do so – even if it is the only way they could avoid the disease. This does not seem to track the ethical reasons at play in this comparison. It seems far more ethically permissible to avoid a lethal disease than avoid a particular eye color. When it comes to avoiding disease, there seems to be little moral difference between using different RGTs. This should be reflected in regulation. What we need then is a consistent approach to the regulation of inheritance, and PGD should serve as the model. Each proposed use of GS and GE should be assessed on a case-by-case basis. When therapies are shown to be safe and effective, they should be added to a list of approved conditions.

1.5 Conclusion

Parents have good reasons to provide benefits for their future children, even when this benefit will change the identity of their future children. There are no reasons why this general principle should not also apply to RGTs. Parents have good reasons to benefit their future children through RGTs, particularly when they can be used to prevent devastating diseases. When it comes to avoiding disease, the choice of RGT is irrelevant – what matters is choosing genes that will benefit your children.

References

Baltimore, David, Paul Berg, Michael Botchan, Dana Carroll, R. Alta Charo, George Church, Jacob E. Corn, George Q. Daley, Jennifer A. Doudna, Marsha Fenner, Henry T. Greely, Martin Jinek, G. Steven Martin, Edward Penhoet, Jennifer Puck, Samuel H. Sternberg, Jonathan S. Weissman, and Keith R. Yamamoto4. 2015. "Biotechnology. A Prudent Path Forward for Genomic Engineering and Germline Gene Modification." Science 348 (6230): 36–38.

Berndt, Sonja I., Stefan Gustafsson, Reedik Mägi, Andrea Ganna, Eleanor Wheeler, Mary F. Feitosa, Anne E. Justice, et al. 2013. "Genome-Wide Meta-Analysis Identifies 11 New Loci for Anthropometric Traits and Provides Insights into Genetic Architecture." Nature Genetics 45 (5): 501–12.

Bocheva, Georgeta, and Nadka Boyadjieva. 2011. "Epigenetic Regulation of Fetal Bone Development and Placental Transfer of Nutrients: Progress for Osteoporosis." Interdisciplinary Toxicology 4 (4).

Bourne, Hannah, Thomas Douglas, and Julian Savulescu. 2012. "Procreative Beneficence and in Vitro Gametogenesis." Monash Bioethics Review 30 (2): 29–48.

Buchanan, Allen E., and Dan W. Brock, eds. 2007. From Chance to Choice: Genetics and Justice. 8th printing. Cambridge: Cambridge Univ. Press.

Callaway, Ewen. 2016. "UK Scientists Gain Licence to Edit Genes in Human Embryos." Nature 530 (7588): 18–18.

Collins, Francis. 2015. "Statement on NIH Funding of Research Using Gene-Editing Technologies in Human Embryos."

Desrivières, S., A. Lourdusamy, C. Tao, R. Toro, T. Jia, E. Loth, L. M. Medina, et al. 2015. "Single Nucleotide Polymorphism in the Neuroplastin Locus Associates with Cortical Thickness and Intellectual Ability in Adolescents." Molecular Psychiatry 20 (2): 263–74.

Hakimi, Parvin, Jianqi Yang, Gemma Casadesus, Duna Massillon, Fatima Tolentino-Silva, Colleen K. Nye, Marco E. Cabrera, David R. Hagen, Christopher B. Utter, Yacoub Baghdy, David H. Johnson, David L. Wilson, John P. Kirwan, Satish C. Kalhan and Richard W. Hanson. 2007. "Overexpression of the Cytosolic Form of Phosphoenolpyruvate Carboxykinase (GTP) in Skeletal Muscle Repatterns Energy Metabolism in the Mouse." Journal of Biological Chemistry 282 (45): 32844–55.

Isasi, Rosario, Erika Kleiderman, and Bartha Maria Knoppers. 2016. "GENETIC TECHNOLOGY REGULATION. Editing Policy to Fit the Genome?" *Science (New York, N.Y.)* 351 (6271): 337–39.

Jang, Kerry L., W. John Livesley, and Philip A. Vemon. 1996. "Heritability of the Big Five Personality Dimensions and Their Facets: A Twin Study." *Journal of Personality* 64 (3): 577–92.

Jinek, Martin, Krzysztof Chylinski, Ines Fonfara, Michael Hauer, Jennifer A. Doudna, and Emmanuelle Charpentier. 2012. "A Programmable Dual-RNA-Guided DNA Endonuclease in Adaptive Bacterial Immunity." *Science* 337 (6096): 816–21.

Lanphier, Edward, Fyodor Urnov, Sarah Ehlen Haecker, Michael Werner, and Joanna Smolenski. 2015. "Don't Edit the Human Germ Line." *Nature* 519 (7544): 410–11.

Liang, Puping, Yanwen Xu, Xiya Zhang, Chenhui Ding, Rui Huang, Zhen Zhang, Jie Lv, Xiaowei Xie, Yuxi Chen,Yujing Li, Ying Sun, Yaofu Bai, Zhou Songyang, Wenbin Ma, Canquan

Zhou, and Junjiu Huang. 2015. "CRISPR/Cas9-Mediated Gene Editing in Human Tripronuclear Zygotes." Protein & Cell 6 (5): 363–72.

Lim, Miranda M., Zuoxin Wang, Daniel E. Olazábal, Xianghui Ren, Ernest F. Terwilliger, and Larry J. Young. 2004. "Enhanced Partner Preference in a Promiscuous Species by Manipulating the Expression of a Single Gene." *Nature* 429 (6993): 754–57.

Magnusdottir, Erna, and M. Azim Surani. 2014. "How to Make a Primordial Germ Cell." *Development* 141 (2): 245–52.

Oikkonen, Jaana, Yungui Huang, Päivi Onkamo, Liisa Ukkola-Vuoti, Pirre Raijas, Kai Karma, Veronica J. Vieland, and Irma Järvelä. 2015. "A Genome-Wide Linkage and Association Study of Musical Aptitude Identifies Loci Containing Genes Related to Inner Ear Development and Neurocognitive Functions." *Molecular Psychiatry* 20 (2): 275–82.

Parfit, Derek. 1992. *Reasons and Persons*. Repr. Oxford: Clarendon.

Sabatello, Maya. 2015. "Regulating Gamete Donation in the U.S.: Ethical, Legal and Social Implications." *Laws* 4 (3): 352–76.

Tang, Yu-Ping, E. Shimizu, G. R. Dube, C. Rampon, G. A. Kerchner, M. Zhuo, G. Liu, and J. Z. Tsien. 1999. "Genetic Enhancement of Learning and Memory in Mice." *Nature* 401 (6748): 63–69.

Theodosiou, Anastasia A., and Martin H. Johnson. 2011. "The Politics of Human Embryo Research and the Motivation to Achieve PGD." *Reproductive BioMedicine Online* 22 (5): 457–71.

Bogdan Olaru

2 Bypassing Morality Through Conventional and Unconventional Forms of Moral Enhancement

2.1 The argument emanating from similarity

After vigorous promotion of moral bioenhancement (Douglas 2008, 2013; Persson and Savulescu 2008, 2013, 2016; DeGrazia 2014), its prospect has encountered much criticism (Harris 2011, 2014, 2016; Schaefer 2015; Focquaert and Schermer 2015). Moral bioenhancement refers broadly to the idea that we should use biomedical means, if available and safe, to extend or supplement the efforts of bettering our moral nature. These new means work directly on the biological level of emotions, motivations, and attitudes. While serving the same aim of ameliorating human interactions from a moral point of view, such direct and unconventional tools are expected to catch up with more traditional and indirect means of moral enhancement, such as education, socialization, parental supervision, wise public policy, as well as classical tools of fostering reason and decision-making, such as advancing knowledge and spreading reliable information. Moral bioenhancement supplements this repertoire of well-established means, helps strengthen morality, and is in the service of a better world. However, one of the most important objections against moral bioenhancement (Harris 2011) is that manipulations of human functioning at the biochemical or neuronal level undercut a person's freedom and moral reasoning. Because of deep-rooted connections between reason, autonomy, and morality, praised by many philosophers, this criticism amounted to exposing a self-defeating feature of any attempt of what might be qualified as moral bioenhancement. If, as the proponents assume, the new envisioned techniques focus on suppressing or increasing the biological layer of emotions, motivations, and/or attitudes to shift the behavioral output in the right direction, the change into a better person seems to occur in ways that are at least dissociated from, if not at odds with, rational scrutiny and moral agency. In other words, enhancing morality through biotechnological means seems to obliterate a hard-to-avoid relation between the idea of morality and moral person on the one hand and reason and justification on the other. As John Harris puts it, "The intervention is designed to bypass reasoning and act directly on attitudes. When such attitudes are manipulated, not only is freedom subverted but also morality is bypassed." (Harris 2014, 372) A fundamental flaw lies at the heart of any attempt at moral bioenhancement: it can only take place in a manner that threatens to erode, generally and in the long term, the very idea of morality. Moral bioenhancement thus falls short of reaching the aim of supporting and safeguarding morality itself.

Is such a threat a new entry in ethics? And does it amount to a clear-cut distinction between conventional (i.e., through education and public policy) and unconventional (mainly via biochemical and neurological input) moral enhancement? John Harris, who acknowledges the distinction (Harris 2011, 104), exposes the risks that might arise with new forms of moral bioenhancement, but these risks are by no means unmatched by the painful, everyday pitfalls that occur during socialization, schooling, and parental education. This observation, which was mostly put forward by supporters of moral bioenhancement, underlines the similarity between new and old methods in order to gain public consent for novel enhancement techniques. For instance, when DeGrazia (2014) tackles the common objection against moral bioenhancement about the content of morality – skepticism that stems from moral pluralism – he argues that similar strain is put on our education system: "One should not inculcate moral values that are wrong, so how can a parent be sure that she or he is justified in providing a particular type of moral instruction? Also facing this challenge are public school teachers who attempt to inculcate in students certain moral virtues such as civility, respect for differences and concern for the poor." (DeGrazia 2014, 363) He recommends that we should stick with what he calls "*points of overlapping consensus among competing, reasonable moral perspectives*" (DeGrazia 2014, 364). The rule holds for both conventional moral enhancement and for whatever biomedical means of moral improvement will be considered appropriate by the state to support, encourage, or even require. There is one notable difference between what parents do when raising their children and what the state can be allowed to promote as better public policy. Parents enjoy more freedom to choose their view about what morality requires, whereas the state must check and approve any proposal of public moral enhancement against the content reached through overlapping consensus. Consequently, it should be easier to accept a public policy about a certain type of moral bioenhancement once the test of "*overlapping consensus among competing, reasonable moral perspectives*" (DeGrazia 2014, 364) is completed. In contrast, because of a broader area of disagreement, which is socially accepted when it comes to raising children, some might never totally get rid of a good portion of emotional discomfort caused by the fact that parents enjoy such significant power over how they morally shape their offspring. Their task is apparently more exposed to failures than a state-driven, robust research based on shared moral values. DeGrazia's argument depends on whether or not it is possible to reach a consensus on substantive matters that could express something vital for each reasonable moral view publicly endorsed in a liberal society and that could satisfy reasonable concerns about other competing views. This matter is far from being settled.

Mark Walker (2009) draws on the same idea of similarity when discussing what he labeled "The Genetic Virtue Project" (GVP) – an interdisciplinary project between philosophers, psychologists, and geneticists aimed at morally improving humankind through biotechnological manipulations of something that he assumes to be "genetic correlates of virtuous behavior." On the one hand, there are good reasons to directly target an individual's behavior, if the aim is to enhance his or her conformity with morality. But, on the other hand, GVP aims at more complex and stable interventions. GVP could emulate through genetic modifications what we actually welcome in parenting: promoting virtues as a way to enhance children's morality. Moreover, the intervention would be so profound that it could be passed on to the next generations. Thus, Walker draws his argument on the analogy between promoting virtue in common ethical practice and the project of genetically selecting for virtue-compatible features: "There is a tendency to think of the virtues being implemented on the knees of parents but this focus, as many have argued, ignores other possible influences on the capacities of individuals to learn to be virtuous. Theorists from Plato and Aristotle to Marx and MacIntyre have emphasized the role of socio-political influences in inculcating virtue. The view that genes may influence how readily humans are able to learn the virtues is meant to join this chorus." (Walker 2009, 43) There is a fundamental convergence between moral enhancement through traditional education and selecting for virtue-promoting genes. The latter supplements parenting and socialization, as well as the parent's moral duty to fostering virtuous behavior. Of course, goals are not all that matter and the dissimilarity of means cannot be overlooked (Focquaert and Schermer 2015). Even if we could one day establish a satisfactory analogy between well-designed genetic screening programs that aims at virtues and socioparental influences that instill virtue in more conventional ways, the analogy will still remain a quite limited strategy to confront the vast amount of criticism brought against moral bioenhancement in terms of safety, side effects, paternalism, individual morality, parochial and opposing moralities, and so on. In contrast, it seems that we are more willing to accept whatever parents do, as moral educators, with their children based on the assumption that they are inspired by their good will and honorable intentions.

However, the argument from similarity has some engaging effect on the debate. It fuels the idea that we should treat nonconventional moral enhancement as part of a process that is both natural and unavoidable. Eventually, we may come to the conclusion that we have to assume a positive moral obligation to (support any research that has a chance to) enhance our lives (Harris 2007, 79, 192). Why should we not extend this obligation to cover the improvement of our moral lives by enhancing moral virtues, emotions, or intuitions? Does morality not enhance our lives? While similarity of aims seems to play into the hands of moral bioenhancement supporters, it should also be the starting point to review what seems rather alarming in all appeals to moral improvement. In the next two sections, I present the alleged risk and then assess the novelty of this risk, which some fear will occur, considering three methods discussed in the literature as plausible candidates for improving peoples' morality. This line of criticism brought against moral bioenhancement relates to the danger that tools designed to improve moral conduct might (1) wrongly select out some ways of thinking and thus distort the very process of finding reasons and justifications, (2) disrupt one's narrative identity to the point of bringing about dangerous internal conflict, and (3) operate fine-tuned mental manipulations of mood and unconscious dispositions that would not have been endorsed by a subject who would have had the chance to properly analyze them. I have chosen these examples to illustrate how moral bioenhancement might compromise one's sense of acting as an autonomous person. However, my intention is to show that their *modus operandi* does not diverge from what people try to achieve through upbringing and conventional education. The argument put forward in the last section explains that precisely this similarity between new and old methods should strike us as a warning signal rather than as an incentive to pursue moral bioenhancement. One should not repeat old and familiar errors.

2.2 Selecting for particular ways of reasoning, disruptions of one's narrative identity, and freedom from any mental manipulations

Alterations of emotions and intuitions are one path to influence moral reasoning and decision-making. But what if special forms of moral bioenhancement would target the reasoning process itself? A great concern, for those who subscribe to the idea that there is a powerful link between reason and morality, is that moral bioenhancement could put at risk autonomous reasoning and the process of justification itself. If the means for such influence were at hand, the subject's integrity could be impaired in a pervasive manner by advocating not necessarily a substantive moral view but rather a particular way of reasoning, i.e., "the narrow conception of reasoning promoted by the enhancer" (Schaffer 2015, 270). To be sure, a moral decision will still be reached through reasoning, yet without knowing whether the reason we think it is crucial was the right reason for the matter at stake or whether the way we reach the conclusion was the soundest deduction one can provide. This is an example of how morality is bypassed by acting directly on reasoning itself. It illustrates how a person could fail without even noticing the need of analyzing his or her decisions and confronting them against the background of the assumptions. The risk is similar to the situation when someone fails to consider relevant features of a moral problem or does not take into account alternative points of view that sustain and help refine moral reasoning. It is of course a matter of pure speculation whether such unconventional tuning of high cognitive functions is possible via biochemical means or electromagnetic brain stimulation.

Another threat posed by the prospect of moral bioenhancement is the danger of causing disruptive interventions on a person's narrative identity (Focquaert and Schermer 2015). Narrative identity is defined as the sum of all features constitutive for how a person understands himself or herself and relates to his or her image. It includes experiences, attitudes, beliefs, values, and desires – in short, it is the cognitive, motivational, and emotional asset of a person. Although we define ourselves through our experiences, attitudes, and beliefs, the construction of one's self does not need to take place in a conscious manner. It is often the rule that we are rather unaware of how deeply rooted our natural inclinations and learned propensities are. We reach a better picture of ourselves simply by self-examination and critical exchange with friends or relatives. We testify the consistence of our self against the assumption that we are at least partially generating central pieces of our identity. Moral bioenhancement has the potential to alter this asset and to induce disrupting identity changes. This line of criticism claims that moral bioenhancement will bring about radical psychological changes that could threaten a person's sense of continuity and coherence, especially when these changes cannot plausibly be felt consistent with experiences that defined the self before the moment when the enhancement occurred. The picture becomes more intricate when such alteration of the self takes place in a quiet, unnoticed manner: "This might lead individuals to unreflectively accept or even welcome certain traits that would not be similarly endorsed by their pre-interventional/pre-enhanced self. (...) It may therefore result in a dissociation between one's implicit narrative self and one's explicit narrative self, that means: in a situation of self-blindness. (...) self-blindness, in this sense, is a form of inauthenticity that threatens the autonomy of the self" (Focquaert and Schermer 2015, 148). Thus, the objection goes on, the enhanced person must be aware of any change that attempts to reframe his or her narrative identity. One must reject interventions that bypass this awareness level as nonethical.

Drawing on a series of studies (Crockett et al. 2008, 2010) that show how serotonin modifies the subjects' moral judgments and curbs the willingness to harm others, Christoph Bublitz (2014, 2016) makes a plea for protecting "freedom of mind" from intrusive "alterations of unconscious dispositions". Freedom of mind is a person's capacity to make use of his or her mental functions and freely dispose of mental states "as she pleases, free from external influences and internal impediments" (Bublitz 2016, 94). In contrast to freedom of will, which is usually linked to the possibility to act otherwise, freedom of mind underlines the requirement of not being subjected to any mental manipulations, such as those described in Crockett's studies. Why would such mental freedom be ethically significant? One detrimental influence of artificially induced motivations, attitudes, or dispositions is that they might put a strain on the person's ability to consciously control his or her own mind. The effect is similar to when some unexpected noise distracts you from what you are just about to do or when you accidentally take a drug overdose with damaging effect on your normal cognitive functioning. It is not the content or the intention that is objectionable in some attempts of altering moral judgment, but rather the fact that, generally, any newly induced mental state has a diverting potential. These interventions simply consume resources and threaten to impair autonomy on the very basic level of accessing and using one's own mental power and capacity.

2.3 Fundamental similarity between conventional and nonconventional moral enhancement

The three examples are taken from a broad and challenging range of criticisms related to moral bioenhancement. They share the concern that some interventions have the potential to undermine the link between morality and autonomous choice guided by reason in ways that could easily go unnoticed while being highly effective. If their potential is confirmed, these interventions might signal the advent of new and very powerful manipulation tools. Yet, they are in fact not so new. They are quite similar to more conventional strategies of promoting morality. In one sense, their novelty comes from the fact that, once a sufficient degree of subtlety and efficiency is attained, unconventional techniques of moral enhancement will achieve a very high resemblance with conventional means of enhancing moral judgment, as listed by some of their advocates: general education (including self-education), knowledge acquisition, introspection and disclosure of our own prejudices, engagement with genuine and urgent problems of the world, developing good policy and legislation, and so on (Harris 2011). The similarity between conventional and nonconventional moral enhancement comes from both the potential of bringing about highly effective changes and the danger of doing it in ways that may contradict morality itself. What often misfires with conventional wisdom and morality can also go wrong with moral bioenhancement, including selective endorsing of specific modes of reasoning, identity disruptions, or unnoticed alterations of one's mental life. In fact, many so-called methods of moral bioenhancement do not seem to threaten liberty, autonomy, and individuality in such a novel and overwhelming manner that no one has already seen happening during well-known stages of moral development and instruction in normal interactions between children and the adults who look after them.

Let us discuss more about this *prima facie* similarity between conventional and nonconventional moral enhancement. Various ways of reasoning are at work each moment we attempt to modify the others' conception on substantive issues. This is happening in everyday discussions on moral questions, as well as in highly abstract philosophical debates. When I attempt to persuade someone, my aim is not only to become the mere vehicle that passes on a substantive moral view. I will put considerable weight on passing on this view along with the right arguments, the best ways of reasoning, and the most productive strategies of backing my claims with facts. These are, as I shall think, the most appropriate means to promote the view I support. I implicitly advocate my "narrow conception of reasoning" while taking it to be the right conception of reasoning meant to secure what I consider to be the right substantive view related to a specific moral issue. Any attempt to improve moral judgment in very simple conventional ways generally promotes its underlying conception of reasoning. So, why should this feature amount to a genuine problem once moral bioenhancement becomes reality, as some critics fear? There is little sense to deplore what might go wrong with reasoning in moral bioenhancement if one fails to acknowledge that the suspected danger, if it really exists, should represent a concern for any other form of moral enhancement. Different forms of moral education might share the same alleged danger of selectively promoting peculiar forms of reasoning. But the willingness to overcome this danger could also be a shared feature. If the latter is true, we should simply focus on lessening the risk of any kind of biases in everyday moral reasoning, as well as in ethical debates. This includes analyzing one's decisions and confronting them against the relevant assumptions, taking into account all relevant features and views, and hoping to gradually sustain and refine moral reasoning.

The other two objections shortly described in the previous section can be considered together. According to the second objection, moral bioenhancement has the potential to severely disrupt personal identity. Artificially induced mental states might threaten a person's sense of continuity and coherence, especially when the identity shift remains undetected. While this threat seems plausible, it is, again, nothing novel. Parents teach their children the values they prize the most. They put effort and great expectation in shaping their children's identity according to what they think to be the right way of living and behave. Later on, grown-up children usually face life experiences that threaten to shake the very identity that their parents strived to polish in their preferred manner. This common test makes them reconsider parts of their most intimate beliefs. Their inner coherence holds and fails with back–and-forth confirmations and refutations of what they believed to be morally right and epistemically true. Cognitive interventions that aim to modify whole systems of beliefs, such as making people more tolerant toward immigrants or more sensitive to the consequences of one's actions in a global world, work as a matter of fact by challenging deep-rooted features of their constructed and/or inherited identity. If sound moral reasoning makes people change their wrong beliefs in a process that fundamentally alters their identity, such alterations can hardly be seen as disruptive. On the contrary, they represent a move toward moral progress, indeed, an example of conventional moral enhancement. How does such enhancement actually work? In most cases, what happens is that people are confronted with facts and affirmations they are not aware of, do not want to deal with, or even intentionally disregard when they are obvious. It often means to make people acknowledge disturbing facts and put a veritable strain on their "freedom of mind", in the sense stated before as freedom from induced mental states that might have a "distracting" potential. Uncovering gross injustice or appalling cruelty can at times be very disturbing and might include and generate by itself undesired "alterations of unconscious dispositions" that challenge one's "peace of mind". Such challenges occur often as a side effect of conscious deliberation on relevant facts and issues, but they sometimes represent the method of choice for shaking a self-serving, closed system of beliefs that seems impermeable to change. In this case, the aim is to destabilize the very inner coherence that bears a protective meaning. Despite its undisclosed way of functioning and its focus on unconscious mechanisms, deliberate erosion of one's innermost freedom of mind is sometimes the only and the right way to make known other people's legitimate motives, interests, and goals. Raising awareness in order to better meet moral requirements that one has perhaps already endorsed is a strategy that overtly aims at disrupting, or at least challenging, one's "peace of mind" by inducing new motivations, attitudes, or dispositions. Such specific cognitive and motivational alterations, of course, consume significant psychological resources. They pose a threat to a person's capacity to freely dispose of his or her mind's integrity and autonomy at a very basic level. The question is whether we can afford to spare us such interventions in the name of a self-sufficient freedom from, as Christoph Bublitz puts it, "external influences and internal impediments".

2.4 The constraint of moral equality between the enhancer and the enhanced person

The purpose of these counterarguments was to show that moral bioenhancement might eventually come close to using the same paths and channeling the same resources that are involved in conventional approaches to moral enhancement. In the view of this potential similarity, it is fair to say that moral bioenhancement – at least in some of its putative forms, such as those described earlier – does not emerge as a novel risk for human autonomy. Of course, some interventions are riskier than others, and time is essential to assess the risk and overcome it. Moral bioenhancement probably has the potential to undermine values such as autonomy and liberty and provide effective means for disastrous social engineering. In other words, it can theoretically undermine or bypass morality itself. However, this cannot be a totally novel risk that takes us by surprise. Lack of critical thinking and information can easily make parenting harmful for a child's psychological integrity, turning parenting itself into a pervasive threat for autonomy later in adulthood. Unwise public policy may turn into undesirable social engineering, too. But the idea that some unconventional ways of moral enhancement (such as those described in the previous sections) are using the same paths and channeling the same resources as conventional ways (such as, let us say, when parents teach their children to be morally good persons) gives hope that same precautions might work in both areas. If this is true, we should not fear that we are not well equipped to counteract a potential danger that seems to be inherent in any attempt to promote morality, be it through conventional or nonconventional ways: the danger that enthusiasm and creative search for new means displayed in the name of moral progress might erode even the most vital moral values. We should feel at least warned by various derailments that have already occurred along so many attempts of social and moral bettering of human persons and communities. One need only think of dubious moralities embedded in cultural practices, from gender-dependent assignment of roles and virtues to discrimination on various grounds. Self-awareness starts with acknowledging that the very act of caring for children and for one's family gives parents a powerful opportunity to choose the ways they think most appropriate to promote their values. Given this opportunity, some may choose to raise their children in ways that neglect how useful rational argumentation and empathy are at critical points in someone's life, making room for intolerance and extremism. One cannot consider this moral enhancement, even if these parents also.

To sum up, the potential danger posed by moral bioenhancement is not at all unparalleled by the danger that already exists when it comes to raising moral conscience through more conventional, e.g., cognitive, means. To be sure, what supposedly might go wrong with moral bioenhancement – the threat of fostering odd preferences for particular frames of reasoning; the prospect of profoundly transforming someone's identity; and the risk of having all kinds of "noising" input flowing into one's mental life – can as well go wrong with conventional moral enhancement, including parental education. A simple rule emerges from this similarity. One should not advocate a specific method designed to morally enhance people's conscience, motives, or behavior without giving enough guarantees that there is a way back to the current state, if the intervention proves to be unsuccessful or undesirable after up-to-date and careful consideration. This possibility is there, at least theoretically, when moral enhancement targets cognitive structures. People often revise the way they have been raised by their parents through concepts and beliefs they acquire at some later stage in life.[1] In contrast, this possibility, which is fundamental to the preservation of freedom and autonomy, is not at all possible for most of the popularized accounts on moral bioenhancement and, in most cases, is not even taken into account (here is an exception: Sparrow 2014). This flaw shows why some accounts are so difficult to defend. Because they focus on causal mechanisms at the biochemical or neuronal level, such accounts are constrained to measure their success in terms of output stability. They do not treat the subjects of potential interventions as dynamic entities but rather as fixed structures that can be made to fit in a devised scheme. The will of the designer prevails over the latent qualities and abilities of the designed subject.

Nevertheless, why should we defend the right to revise and invalidate an intervention that makes changes for the better even when we could reach absolute confidence that the change is beneficial for the subject (similar to so many parental influences that bring about positive changes in children)? Part of the answer has to do

---

1 This seems to me the kind of statement anyone would find best illustrated by examples taken from his or her own experience. Yet, if we do not want to get caught in the deep waters of any old and new empirical research, one will find similar insights from rather unexpected intersections of theoretical accounts. Drawing on Freud's views on parental authority and its role in moral learning, John Rawls agrees that "since parents and others in authority are bound to be in various ways misguided and self-seeking in their use of praise and blame, and rewards and punishments generally, our earlier and unexamined moral attitudes are likely to be in important respects irrational and without justification. Moral advance in later life consists partly in correcting these attitudes in the light of whatever principles we finally acknowledge to be sound." (Rawls 1999, 402)

with preserving freedom and autonomy. It is also a question of moral equality. It might be useful to recall in this context a similar worry that emerged about 2 decades ago around reprogenetics and positive eugenics. The prospect of combining reproductive and genetic technologies in order to single out desired physical or psychological traits raised concerns that more and more would-be parents will be tempted to genetically enhance their offspring. Of course, many projections extensively discussed in the literature remain even today pure possibilities. However, the ethical debate captured an essential aspect touching core values in a liberal society: the idea that people form a moral community of free and equal persons, an idea that spans a wide tradition of liberal thinking, starting with John Locke and reaching its highlight with John Rawls. In a 2001 lecture, Jürgen Habermas worked out an argument against liberal eugenics. It reads as follows: Any parent–child relationship is inevitably asymmetrical. In normal socialization processes, parents exercise a vast influence over their children. A parent's goals and expectations, as well as the means chosen to fulfill them, do sometimes match and sometimes do not match future self-assumed intentions, goals, means, and expectations of grown-up children. Still, the risk of dissonant cases is compensated by the fact that whatever influence parents exert over their offspring, this influence is essentially open to question. As Habermas put it, children "can retrospectively compensate for the asymmetry of filial dependency by liberating themselves through a critical reappraisal of the genesis of such restrictive socialization processes" (Habermas 2003, 62). By critically reassessing the life projects their parents specially devised for them, children acquire the status of free and equal persons transgressing intergenerational boundaries.

However, reassessment of goals and means is no longer possible for people selected via genetic programming for some specific traits and dispositions:

> Eugenic interventions aiming at enhancement reduce ethical freedom insofar as they tie down the person concerned to rejected, but irreversible intentions of third parties, barring him from the spontaneous self-perception of being the undivided author of his own life. [...] A universalistic understanding of law and morality rests on the assumption that there is no definite obstacle to egalitarian interpersonal relations. [...] No dependence on another person must be irreversible. [...] Eugenic programming establishes a permanent dependence between persons who know that one of them is principally barred from changing social places with the other. But this kind of social dependence, which is irreversible because it was established by ascription, is foreign to the reciprocal and symmetrical relations of mutual recognition proper to a moral and legal community of free and equal persons. (Habermas 2003, 63–5)

The asymmetrical relationship between an enhancer and the enhanced is obvious in both trait selection via genetic programming and moral enhancement through different methods that aim at modulating emotions, attitudes, or behavior. One does not need to embrace Habermas's account of communicative action as a foundation of any interaction in order to see how basic human relationships might irreversibly shift. However, these observations make a plea for disclosing what normative concept of human person and humanity one chooses when it comes to assessing whether the shift takes place in the right direction or not. For Habermas, this normative concept is the idea of free and equal persons.

Here is an illustration from another field. When John Rawls describes what psychological constraints should form the basis for assessing the reasonableness of a theory of justice, he embarks on his favorite constructivist approach and outlines a moral psychology that fits the task. He constructs persons through the lens of their two moral powers. These are assumptions that lead to the preferred concept of person around the idea of moral equality. No reasonable conception of justice can bypass this concept and overlook the fact that people enjoy equal moral status. To hold on the concept of moral equality is a normative assumption. As Rawls (1993, 87) put it in *Political Liberalism*: "Human nature and its natural psychology are permissive: they may limit the viable conceptions of persons and ideals of citizenship, and the moral psychologies that may support them, but do not dictate the one we must adopt."

Now, any advocate of moral bioenhancement should assess the recommended method from a similar line of reasoning. The analogy is straightforward. The first question one should ask is: Does this particular method of morally enhancing people's intentions, emotions, attitudes, or behavior have the potential of asymmetrically shifting the relationship between free and equal persons? The second question would be: Is there any guarantee that the designed intervention to morally enhance people's intentions, emotions, and so on can be critically revised and undone if it proves unacceptable from the subject's perspective? This question is important as a check against the fundamental asymmetry between the enhancer and the enhanced subject. The enhancer must ensure that the person enhanced can reverse the intervention, if he or she finds it wrong. The second question cannot be dismissed by showing that the proposed intervention is morally good beyond any doubt and that no one will ever be able to consider it unacceptable. Their status as free and equal persons is at stake before and after the intervention. Both parties agree that it is a requisite for a moral life to preserve their reciprocal recognition as free and equal members of a moral community. Should moral bioenhancement generate a long-lasting imbalance detrimental for this kind of recognition between free and equal persons, it will compromise morality itself. Traditional means of moral enhancement are not essentially different compared to unconventional means from the perspective captured in the first question. Various forms of raising awareness for any kind of moral values and goals, work often, whether during childhood or later in life, through persuasion, rational constraints, and generally by guiding the moral subject. They could sometimes amount to a veritable sort of "maneuvering" of the subject toward the correct belief or behavior. Such interventions are, to a certain degree, similar to objectifying a person and his or her will to a greater aim. Thus, it will be a mistake not to see that nearly any common interactions between an educator and his or her pupil bear the potential of bringing about the same imbalance concerning their status as free and equal persons we fear from unconventional moral enhancement. Here too, it is true that preserving moral equality is a requisite for a moral life.

References

Bublitz, Christoph. 2016. "Moral Enhancement and Mental Freedom." *Journal of Applied Philosophy* 33(1): 88-106.

DeGrazia, David. 2014. "Moral Enhancement, Freedom, and What We (Should) Value in Moral Behavior." *Journal of Medical Ethics* 40(6): 361-368.

Douglas, Thomas. 2008. "Moral Enhancement." *Journal of Applied Philosophy* 25(3): 228-245.

Douglas, Thomas. 2013. "Moral Enhancement via Direct Emotion Modulation: A Reply to John Harris." *Bioethics* 27(3): 160-168.

Focquaert, Farah, and Maartje Schermer. 2015. "Moral Enhancement: Do Means Matter Morally?" *Neuroethics* 8(2): 139–151.

Habermas, Jürgen. 2003. *The Future of Human Nature*. Cambridge: Polity Press.

Harris, John. 2007. *Enhancing Evolution. The Ethical Case for Making Better People*. Princeton, NJ: Princeton University Press.

Harris, John. 2011. "Moral Enhancement and Freedom." *Bioethics* 25(2): 102-111.

Harris, John. 2014. "Taking Liberties with Free Fall." *Journal of Medical Ethics* 40(6): 371-374.

Harris, John. 2016. "Moral Blindness – The Gift of the God Machine." *Neuroethics* 9(3): 269-273.

Persson, Ingmar, and Savulescu, Julian. 2008. "The Perils of Cognitive Enhancement and the Urgent Imperative to Enhance the Moral Character of Humanity." *Journal of Applied Philosophy* 25(3): 162-177.

Persson, Ingmar, and Julian Savulescu. 2013. "Getting Moral Enhancement Right: The Desirability of Moral Bioenhancement." *Bioethics* 27(3): 124-131.

Persson, Ingmar, and Julian Savulescu. 2016. "Moral Bioenhancement, Freedom and Reason." *Neuroethics* 9(3): 263-268.

Rawls, John. 1993. *Political Liberalism*, 2nd ed. 1996. Cambridge, New York: Columbia University Press.

Rawls, John. 1999. *A Theory of Justice*. Rev. ed. Cambridge, Massachusetts: Belknap Press.

Schaefer, G. Owen. 2015. "Direct vs. Indirect Moral Enhancement." *Kennedy Institute of Ethics Journal* 25(3): 261-289.

Sparrow, Robert. 2014. "Better Living Through Chemistry? A Reply to Savulescu and Persson on 'Moral Enhancement'." *Journal of Applied Philosophy* 31(1): 23-32.

Walker, Mark. 2009. "Enhancing Genetic Virtue: A Project for Twenty-first Century Humanity?" *Politics and the Life Sciences* 28(2): 27-47.

Marcello Ienca

3 The Neuroenhancement Continuum and the Minimal Rule

3.1 Introduction

This chapter argues that pharmacological nootropics should not be seen, from an ethical standpoint, as qualitatively different from other activities such as healthy diet, sleep, education, mental and bodily exercise, parental care, and information technology. In contrast, it suggests that nootropic-induced enhancement should be viewed in continuity with other nonpharmacological activities through which "our uniquely innovative species tries to improve itself" (Greely et al. 2008). My argument proceeds as follows: first, I provide an overview of various forms of neuroenhancement and discuss their impact on cognitive functioning; second, I argue that forms of enhancement that have similar effects on cognitive functioning should have similar levels of ethical permissibility regardless of the physical medium through which they are administered. Since nootropics and other neuroenhancers have qualitatively analogous effects on cognitive functioning, I conclude that they should be seen as part of the same moral continuum.

Of course, claiming that different classes of neuroenhancers belong to the same functional and moral category does not imply that they are equivalent in every respect. It is reasonable to object, for instance, that enhancement by nutrition and nootropics is metabolically different than, say, mental exercise, as it involves a change in what we ingest. However, such intercategorical differences are equally distributed across the whole neuroenhancement continuum and do not justify the dichotomy between nootropics vs nonnootropic enhancers. In other words, such differences do not provide sufficient ground for justifying *a priori* (i.e., based on theoretical deduction rather than empirical observation) contrasting moral judgments between nootropics and all other neuroenhancers.

My argument proceeds as follows. In order to justify *a priori* contrasting moral judgments between the class of nootropics (N) and that of all other enhancers (OEs), there must be at least one morally relevant property P^1 that is peculiar to N, i.e., which is a property of N without being a property of OE as well. But there is no morally relevant property P^1 peculiar to N. Therefore, a contrasting moral judgment between N and OE cannot be justified *a priori*. And consequently, N and OE are *a priori* morally equivalent. In addition, I propose a minimal rule for the administration of neuroenhancing drugs and test the application of this rule in relation to a number of nootropics. This rule aims at providing guidance to physicians during the prescription of nootropics, as well as to policy bodies during the regulation of neuroenhancers.

3.2 Nootropics and the origin of human neuroenhancement

The history of neuroenhancement dates back to the origin of human civilization. Written evidences and archeological tests prove that, at least since the beginning of the Bronze Age, many different cultures developed strategies to augment or extend some core mental capacities. In premodern societies, enhancement was usually obtained through both exercise techniques and the consumption of natural extracts of neuroenhancing substances, such as *Ginkgo biloba*, ginseng, cocaine, guaranà, yerba maté, coffee, yaupon holly, and many others (Lloyd 1911). It is also known that some epoch-making advances in the biological and cultural evolution of the human species, such as the origin of verbal language and tool use, the achievement of a metabolically adequate nutritional regime, the invention of writing, and the institution of education systems for children, exerted a massive positive feedback on our mind's functioning (Barkow et al. 1995). Nevertheless, ethical concerns about neuroenhancement have arisen only in the past 2 decades after some pharmaceutic psychostimulant drugs, chiefly analeptics such as Ritalin and Modafinil, have been invented and widely marketed.

The term *nootropic* (from the Greek νοῦς "mind" and τρέpein "to turn") was coined in 1972 by the Romanian psychologist and chemist Corneliu E. Giurgea to refer to the whole set of pharmaceutical drugs, supplements, nutraceuticals, and functional foods that improve processes such as attention, memory, concentration, intelligence, motivation, perception, and decision-making (Giurgea 1973).

Nootropics usually work in three possible ways: (i) by altering the availability of the brain's supply of neurochemicals, such as neurotransmitters, hormones, and enzymes; (ii) by stimulating nerve growth; and (iii) by increasing the brain's oxygen supply. In conformity with this variation in the way they influence our nervous system, we can branch nootropics into three main families: (A) neurochemical suppliers, (B) nerve growth enhancers, (C) antioxidants and neuroprotectives. Standard classifications further subdivide these three groups into subcategories in conformity with either their cellular function or their chemical composition (Table 1).

As Table 1 shows, most nootropics fall under the first category. This depends on the fact that the most direct way for a given substance to influence neuronal functionality is affecting neurotransmitters or the components of the nervous system that use such neurotransmitters. Cholinergics, for example are substances that affect the neurotransmitter acetylcholine (a facilitator of memory formation) or components of the nervous system that use acetylcholine. Analogously, dopaminergics, the most popular subfamily of neuroenhancers (e.g., L-phenylalanine, amineptine, and methylphenidate), affect the neurotransmitter dopamine or the components of the nervous system that use dopamine. Other substances affect cognition *ex negativo*: rather than supplying a certain neurotransmitter, they decrease the inhibiting neurotransmitter's release. The H3-receptor, for instance, decreases the release of inhibiting transmitters, such as histamine, norepinephrine, and serotonin, and thus increases cognition and wakefulness. Nerve growth enhancers increase brain communication, making brain processing faster. Antioxidants prevent neural oxidative stress, thus inhibiting neuronal aging and death. Neuroprotectives protect neurons from apoptosis or degeneration, thus exerting a preservative effect against brain injury and neurodegenerative diseases of the central nervous system such as Parkinson's, Alzheimer's, schizophrenia, and stroke or by simply preserving cognition from mental deterioration.

Table 1. Overview of Nootropics

Neurochemical suppliers	Nerve growth enhancers	Antioxidants & neuroprotectives
• **Cholinergics:** Choline, OMAE, Meclofenoxate, Huperzine A, Oonepezil, • **Serotonergics:** 5-HTP, Trytophan, Pyiridoxal • **Dopaminercics:** L-Phenylanine, Biopterin, Amineptine, Bupropion, Methylphenidate, MAO-B inhibitors, Seligeline • **Histamine antagonists:** Ciproxifan, A- 349, A-821, ABT-239 • **Amphetamines:** Adderall, Oexedrine, Methamphetamine • **Adrenercics:** Atomoxetine, Roboxetine, Synephrine • **Xanthines:** Caffeine • **Euceroics:** Adrafinil • **Direct Hormones:** Vasopressin Orexin	• **Growth Enhancers:** Lion's Maine Mushroom, lnositol	• **Simple Antioxidants:** ldebenone, Melatonin, Coenzyme Q-10 • **Chief Antioxidants:** Glutathione • **Precursors to Antioxidants:** Acetylcysteine (L-Cystein) • **Neuroprotectives:** Acetylcarnitine, Apoaequorin

The most famous nootropics currently on the market are Ritalin (methylphenidate) and Adderall (amphetamine salt-based psychostimulant). These drugs are used primarily with therapeutic aims to treat people with cognitive difficulties such as Alzheimer's disease, Parkinson's disease, and attention deficit hyperactivity disorder (ADHD).

However, more widespread use is being found by some cross-sectional surveys, because many of the pharmaceuticals used to treat psychiatric and neuropathological conditions have also turned out to improve the performance of the healthy. In 2005, one famous survey estimated that almost 7% of students in US universities have used Ritalin and Adderall in this way, and that on some campuses, up to 25% of students had used them in the past year (McCabe et al. 2005). In Switzerland, a recent survey has revealed that 13.8% of Swiss students have used prescription or illicit nootropics to increase their learning and cognitive performance (Maier et al. 2013).

Due to their effects on the catecholamine system, these classes of nootropics increase executive functions in patients and healthy normal people, improving their abilities to concentrate, manipulate information in working memory, and make it available for further information processing. Since its introduction in 2001, a newer drug, modafinil (an acetamide analeptic), has also shown enhancement potential. The drug was approved by the U.S. Food and Drug Administration (FDA) for the treatment of narcolepsy, shift work sleep disorder, and excessive daytime sleepiness associated with obstructive apnea. Modafinil has become very popular among healthy university students in Western Europe, especially in the U.K. While some students obtain the drug illicitly (diversion of prescribed medication), others use online pharmacies (Coveney 2011).

Attention, concentration, processing speed, and accuracy are not the only mental capacities that can be enhanced through nootropics. A modest degree of memory enhancement is also possible with the ADHD medications just mentioned, as well as with drugs developed for the treatment of Alzheimer's disease such as Aricept (Donepezil), which increase levels of acetylcholine in the brain. Many other compounds with different pharmacological actions are in early clinical trials, having shown positive effects on memory in healthy research subjects.

Pharmacological nootropics have raised massive ethical concerns and sparked debate on their moral and legal permissibility (Dees 2007; Racine and Forlini 2010; Meyers 2014). Despite this disproportionate ethical attention on pharmacological products, not all nootropics have been synthetized in the laboratory. Omega-3 fatty acids, for instance, which increase cognitive performances and decrease the risk of impairment of cognitive speed (Fontani et al. 2005), are fats commonly found in marine organisms and plant oils. Isoflavones, which are related to cognitive function, are produced almost exclusively by vegetables of the Fabaceae (i.e., *Leguminosae*, or bean) family. The same goes for vitamin compounds such as B vitamins, which influence cognitive function through an effect on levels of methylation and homocysteine, and vitamin D, whose active form seems to be involved in brain development and adult brain function.

3.3 The neuroenhancement continuum: nootropics and other enhancers

Brain functioning can be improved not only by means of pharmaceuticals but also by nontherapeutic tools such as adequate exercise, nutrition, parental and medical care, and sleep. It is well known that all these kinds of activities can massively influence the functioning of our mind and that their deprivation causes permanent damage on the development of cognitive faculties. Nutrition is a paradigmatic example: evidence shows that undernutrition, especially at an early age, affects brain growth and intelligence quotient (IQ). For instance, the majority of students with the lowest scholastic achievement scores present suboptimal head circumference (anthropometric indicator of past nutrition and brain development) and brain size (Leiva et al. 2001). Recently described effects of dietary factors on neuronal function and synaptic plasticity have revealed some of the vital mechanisms that are responsible for the action of diet on brain health and mental function, therefore illuminating – at the molecular level – this causal relation subsisting between food and cognition. In particular, research in the nutrition sciences has provided exciting evidence for the influence of dietary factors on specific molecular systems and mechanisms that maintain mental function. For instance, a diet that is rich in omega-3 fatty acids is garnering appreciation for supporting cognitive processes in humans and upregulating genes that are important for maintaining synaptic function and plasticity in both humans and rodents (Fontani et al. 2005). On the other hand, diets that are high in saturated fats, common in junk food, are becoming notorious for reducing molecular substrates that support cognitive processing and increasing the risk of neurological dysfunction in both humans and nonhuman animals (Wu and Miller 2005). Similar conclusions can be drawn for sleep, which is needed to regenerate many cellular components of the body, especially the brain, so that they may continue to function optimally. After periods of extended wakefulness or reduced sleep, neurons may begin to malfunction, visibly affecting a person's behavior. In particular, certain stages of sleep are needed for the regeneration of neurons and glial cells within the cerebral cortex, while other stages of sleep seem to be used for forming new memories and generating new synaptic connections. The effects of sleep deprivation on behavior have been tested in relation to the presence of activity in different sections of the cerebral cortex. Results show that some brain areas can be heavily damaged from bad sleep habits. During verbal learning tests on subjects who are fully rested, for instance, functional magnetic resonance imaging (fMRI) scans show that the temporal lobe of the cerebral cortex – i.e., a brain area associated with the processing of language – is very active. However, in sleep-deprived subjects (daily sleeping hours <5), there is no activity within this region (Chee and Chuah 2008).

Research on brain development shows that similar positive effects on cognition can be attributed to education (Brayne et al. 2010), mental and physical exercise (Hillman et al. 2008), the use of information technologies (Bonavita et al. 2015), and parental care during infancy (Feldman et al. 2014).

3.4 The neuroenhancement continuum II: cognition and other systems

Immunology constantly seeks to enhance people's immune system in order to increase their protection against disease. In accomplishing this task, physicians make use of three classes of measures: preventive clinical measures (e.g., vaccines), therapies (e.g., antibiotics), and the promotion of nonclinical enhancers (lactobacilli, vitamin C, and so on). All these measures involve significant alteration of the original system functions or architecture. For example, most people at birth are not immune to diseases such as hepatitis B, poliomyelitis, and rubella. However, in several countries, compulsory vaccination policies have enhanced baseline immunity and augmented individual and collective defenses against communicable disease. Antibiotics can rapidly cure many diseases that otherwise could lead to the patient's death. Nontherapeutic enhancers present in food and supplements can power the system and allow its better functioning. In this process of enhancing the immune system, physicians and public health promoters do not hold the assumption of a presumed default mode of the system. They rather aim at system optimization, namely, the state of the immune system in which it is most capable to protect an organism against disease. In other words, they try to indefinitely maximize the functioning of that system. The only clause that is usually put to this indefinite enhancement of the immune system is that the system augmentation should not thereby cause negative effects of comparable relevance on other related systems. The same thing can be said of the enhancement of the cardiovascular (indefinite optimization through preventive nutrients, pharmaceuticals, exercise, and implants) and the locomotion system (indefinite optimization through preventive nutrients, drugs, exercise, prosthetic limbs, and means of transport).

This paper argues that this same implicit decision-making strategy adopted by immunologists, cardiologists, and locomotion researchers should be applicable to neuroenhancement. In contrast to dichotomist classifications that attribute qualitatively different moral statuses to nootropics and nonpharmacological enhancers, this article proposes to conceptualize the various forms of neuroenhancement as a *continuum*.

Objections against the neuroenhancement continuum often rely on an alleged moral asymmetry between two loosely defined qualities: the *artificial* and the *natural*. Traditional cognition enhancers are seen as *natural*, whereas nootropics are often perceived as *artificial* products of laboratory research. However, the appeal to the *natural* as a morally relevant category is vague. As for an appeal to the *artificial*, the lives of almost all living humans of our time are profoundly unnatural. We live in human-built homes, wear manufactured clothes, manipulate cultivations and farming, and mediate most of our activities via technology. Furthermore, we enjoy medical care, whose invention and development have helped our species in solving many of its adaptive problems. All these aspects of our life bear very little relation to our species's *natural* state, however defined. To this extent, all cognition-enhancing tools with widespread moral acceptance (education, the Internet, writing, portable computers, healthy diet, mental training, and so on) are, broadly speaking, *artificial*. Even in the case of nutrition, which is intuitively the best candidate for *naturality*, most cognition-enhancing nutrients are not directly consumed in their *natural* form. Vitamins, for instance, are frequently consumed as supplements or *artificially* added to drinks and food. Similarly, iodine, which is the most important deterrent of mental retardation, is mostly available as an additive in cooking salt. Therefore, artificiality is not peculiar to nootropics. Saying nootropics should not be morally permissible because they are not natural would thus imply that several socially accepted substances and activities should be considered impermissible, ceteris paribus.

In the absence of solid *a priori* justification, neuroenhancers should be assessed through an evidence-based approach. This approach is discussed in the next section.

3.5 A minimal rule for the administration of neuroenhancers: MiRNA

Considering the moral equivalence between nootropics and traditional enhancers, as well as the parity principle between neuroenhancement and the enhancement of other systems of the human body, it seems unjustified to qualitatively differentiate *a priori* between the moral status of nootropics and that of other neuroenhancers. In contrast, I call for an evidence-based approach to the risk–benefit ratio of neuroenhancers. This ratio can be expressed in the form of a simple heuristic for supporting moral decision-making, which I call the Minimal Rule for Neuroenhancement Administration (MiRNA):

> If something can improve one or many mental abilities, without thereby causing side effects of comparable relevance, then it is morally permissible to promote its application and diffusion.

MiRNA has three advantages. First, it allows a shift from *a priori* justification to evidence-based evaluation. Second, MiRNA offers a simple operational model for moral judgment concerning neuroenhancement. The rule can be expressed as follows:

$$\mathrm{n} \in \{\mathrm{mpN\ iff\ n} = b/c > 1\}$$

where n is every given nootropic about which we want to make a moral judgment, b is the variable computing the potential benefits of this nootropic, c is the variable computing its negative side effects (costs), and *mpN* is the set of the morally permissible nootropics.

This set can be consequently described as follows:

$$\mathrm{mpN} = \{\forall \mathrm{n}\ (b/c > 1)\}.$$

Third, as one can infer from its structural generality, MiRNA has the advantage of not being restricted to the sole nootropics domain, but it is rather applicable to all neuroenhancers. In this case, n should be replaced with e (enhancer), and the set *mpN* with the set *mpE*, whose description is $\mathrm{mpE} = \{\forall \mathrm{e}\ (b/c > 1)\}$. MiRNA might provide a useful framework to guide ethical decision-making in the context of an increasing variety of neuroenhancing solutions and techniques.

3.6 Application of the MiRNA

MiRNA permits the indexing of nootropics and other enhancers under different categories in conformity with their degree of moral permissibility. For the general aim of this paper, three moral categories of nootropics can be distinguished: (A) permissible without restrictions, (B) permissible with restrictions, and (C) impermissible unless exceptions apply. However, it is worth remembering that the results of the application of MiRNA have a degree of variability dependent on the nootropic consumer. Therefore, it may happen that the same drug is permissible for a certain class of consumers under certain conditions and not permissible for a different one under different conditions. Note that such a degree of variability does not make MiRNA vacuous. Rather, it is a characteristic of substance administration policies. Take for instance alcohol. In most legislation systems, alcohol consumption is regulated by consumer-dependent and context-dependent variables: in the US, for instance, alcohol consumption is permissible only for subjects above the age of 21. Other variables concern the place of consumption (permissible at home or in the bar, not permissible in the workplace or in the street) and the activity during which the substance is consumed (permissible when related risks are minimized, not permissible when related risks are high such as say while driving a car).

Considering this degree of consumer- and context-dependent variability, it seems highly reasonable to index in Category A all those nootropics whose use and administration have been scientifically proven to result in potential benefits and/or cause negligible adverse effects. Substances that can be indexed under this category are iodine, many natural adaptogenics and stimulants (caffeine, beta-blockers, *Bacopa monnieri*), omega-3 fatty acids, isoflavones (daidzein, genistein, and glycitein), cholinergics (arecoline), some acetylcholinesterase inhibitors (sage, rosemary), some vasodilators (*Ginkgo biloba*), some reuptake inhibitors (coluracetam, ginsenoside sources), some nerve growth stimulators (melatonin, glutathione), and all vitamins. Nootropics indexed in this category typically display demonstrated potential benefits and have been scientifically proven by laboratorial and epidemiological studies to effectively enhance cognition. Benefits range over a wide variety of enhancements, such as neuronal antioxidation (glutathione, melatonin, rosemary), improvement in short-term memory performance (*Ginkgo biloba*, caffeine, melatonin, *Bacopa monnieri*), long-term memory potentiation (melatonin, vitamin D, caffeine, pramiracetam), IQ improvement (omega-3, iodine), slowed rate of brain atrophy (B vitamins), cognitive speed (omega-3, caffeine), improvement in spatial working memory (isoflavones, vitamin D), improvement in visual–spatial memory and construction (isoflavones), learning (arecoline, omega-3, vitamins B and D), selective attention (*Ginkgo biloba*), and verbal fluency (isoflavones). In addition, these nootropics are typically characterized by having very low side effects, whose potential harm levels go from the relatively harmless (omega-3, *Ginkgo biloba*, sage, vitamins) to the moderately harmful adverse effects of little clinical importance. Reported side effects of Category A nootropics include somnolence (melatonin), insomnia (ginseng, caffeine), allergic reactions (rosemary), and hyperthyroidism (iodine). The characteristic of Category A nootropics is that these side effects are epidemiologically infrequent (affect not more than 2%–3% of patients) and tendentially related to substance abuse (not observed in consumption of small doses).

Under Category B are indexed all those nootropics whose administration has been scientifically proven to provide high potential benefits with an appreciable degree of unintended negative side effects. Substances that can be indexed under this category are eugeroics (modafinil, adrafinil), adrenergics (atomoxetine, reboxetine, synephrine), some dopaminergics (methylphenidate), some a2A receptor agonists (guanfacine), direct hormones (vasopressin, pregnenolone, orexin), fipexide, some racetams (oxiracetam and aniracetam), and the acetylcholinesterase inhibitor huperzine. These nootropics typically display high potential benefits and have been scientifically proven by many laboratorial and epidemiological studies to effectively enhance cognition (Turner et al. 2004). Benefits range over a wide variety of enhancements, such as enhancements of attention (oxiracetam, methylphenidate), general cognitive function (huperzine), concentration (modafinil, neuroleptics, methylphenidate), alertness (modafinil, methylphenidate), memory (oxiracetam, donepezil, pregnenolone), and scores in tests for logical performance (oxiracetam). Category B nootropics are typically characterized by having appreciable adverse effects, whose potential harm levels range between the moderate and the considerable, but still widely below the benefit levels. Reported side effects of Category B nootropics include nausea and vomiting symptoms (donepezil, nicergoline), dizziness and drowsiness (guanfacine, nicergoline, modafinil), headache (guanfacine), nervousness (methylphenidate, modafinil), and anxiety (modafinil). These side effects usually have an epidemiological rate between 2% and 13% of cases and can appear even at normal dosage.

Under Category C are indexed all those nootropics that are not scientifically proven to confer significant benefits, and/or cause high and/or disproportionate adverse effects. Nootropics indexed in this category typically do not fulfill the condition $b/c>1$. Common Category C nootropics are stimulant alkaloids (cocaine), nicotine, fipexide, empathogen–enactogen drugs, benzodiapezines (Restoril; Normison; Euhypnos), dissociatives (phencyclidine [PCP]), and barbiturates (Nembutal, Seconal, Amytal). These agents typically either have no experimentally significant positive effects or do have little positive effects disproportionate to the adverse effects. Reported side effects of Category C nootropics include fever (fipexide), hepatitis (fipexide), cardiovascular disease (heroin, dehydroepiandrosterone [DHEA]), Olney's brain lesion (PCP), coma (Nembutal, Euhypnos), and respiratory depression (Euhypnos, Normison). All these adverse effects could lead to death. In conformity with MiRNA, Category C nootropics, as they fail to fulfill the condition $b/c>1$, should not be considered morally permissible for unrestricted administration among healthy subjects.

3.7 Safety, self-determination, equality, and information

Approaching the problem of cognitive enhancement from the perspective of risk–benefit ratio does not imply giving up substantive ethical concerns. On the contrary, it involves at least three substantive ethical concerns. The first concern is safety. Nootropics must not only be effective but also safe for consumers. This concern is synthetically expressed by the principle that the maximization of potential benefits must be balanced against the minimization of adverse side effects. For newly developed nootropics, two conditions must be fulfilled to ensure safety and thus minimize side effects: (i) all phases of experimental trials must be conducted wherever possible during clinical testing; and (ii) unregulated off-label administration and marketing must be prevented. Clinical trials involving new drugs are commonly classified into four phases:

1. Screening for safety, i.e., the phase in which researchers test an experimental drug or treatment on a small group of subjects to evaluate its safety, determine a safe dosage range, and identify negative side effects.
2. Establishment of the testing protocol, i.e., the phase in which the experimental study drug or treatment is extended to a larger group of subjects to further evaluate its safety and to test its effectiveness on a large scale.
3. Final testing, i.e., the phase in which the experimental study drug or treatment is given to large groups of subjects to definitively confirm its effectiveness, monitor the identified side effects, compare it to treatments already in use, and collect information that will allow safe use of the drug (or treatment).
4. Postapproval studies, i.e., the postapproval and postmarketing phase in which the risks and benefits are further monitored and the drug's optimal use is finally delineated.

For most nootropics on the market today, this fourth phase has not been fully completed. Therefore, further longitudinal studies focusing on the long-term side effects are still required. Until no conclusive results concerning long-term experimental trials are available, no precise quantification of the drug's potential side effects is possible. And, consequently, no clear benefits–costs ratio can be computed. In the absence of such evidence, the applicability of MiRNA remains limited. In fact, as long as the long-term side effects of a given nootropic *n* are partly unknown, there is no certainty that the upfront benefits override the long-term costs.

It is worth pointing out, however, that a drug that causes serious adverse medical consequences but restores good cognitive functioning in severely ill individuals (i.e., people with advanced dementia) might be deemed safe enough to prescribe. Nonetheless, these risks would be unacceptable for healthy individuals seeking enhancement, in particular, for children. In these cases, adjustments to the parameters might be required. For example, for all cases where the threshold of risk toleration must be low or very low (as in the case of healthy children), we need to (i) set the value of b/c as much greater than 1; and (ii) specify the value of c as close to zero. The resulting rule would then be as follows:

$$n \in \{mpM\} \;\; iff \; n = \frac{b}{c} \gg 1\,, with\, c \cong 0$$

The second concern is self-determination. Appropriate ethical guidelines should safeguard the rights to self-determination, mental integrity, and cognitive liberty of individuals (Ienca and Andorno 2017). With the possible exception of specific occupational figures (e.g., surgeons and airplane pilots), people must reserve the right to refute pharmaceutical enhancement. In a specular way, people willing to profit from cognitive enhancement must have the right to do it, provided (i) the nootropic in question meets all safety requirements, (ii) the subject in question is psychologically and legally responsible for his or her own decisions, hence entitled to exercise his or her right to cognitive liberty. To this extent, a special case is represented by children, dementia patients, and in general individuals who cannot be considered mentally competent. In such cases, the subject's deliberation must be balanced by a careful assessment of the medical authority, advanced directives, and legal proxies.

The third concern is equality. Considering that employers, schools, or governments should not generally require the use of cognitive enhancements in order to safeguard the right to self-determination and prevent indirect coercion, they must nevertheless guarantee that neuroenhancers are fairly distributed. Appropriate policy governing the use of cognitive enhancement should therefore avoid exacerbating socioeconomic inequalities, in addition to making nootropics as cheap and capillary distributed as possible.

The fourth concern is information. Appropriate policy should seek to increase public understanding of neuroenhancement and broadly disseminate information concerning the risks, benefits, and eventual alternatives to pharmaceutical neuroenhancement. Alternatives might include healthy nutrition, regular exercise, optimization of sleep patterns through self-monitoring, and other options in the neuroenhancement continuum. In addition, an examination of the social values and pressures (in particular, at school, in academia, and in the business and finance world) that make cognitive enhancement so attractive is highly required. This information would be provided by physicians, psychotherapists, teachers, and college health centers, similar to the way that information about healthy diet, recreational drugs, and sexually transmitted diseases is now disseminated.

3.8 Conclusion

This paper has argued that nootropics present no morally relevant intrinsic properties that differentiate them from other forms of enhancement, in the force of which one can (a) split the whole spectrum of neuroenhancement between nootropics vs other enhancers, and (b) justify *a priori* differential moral judgments. This drive for *a priori* justifications of the moral impermissibility of nootropics might not rest on epistemological but rather on psychological grounds. Indeed, the *a priori* refusal of nootropics might be similar to the general opposition to new techniques and technologies, a phenomenon typically prevalent during the first phases of development of such technologies. Historically observed, several techniques or technologies considered well established currently, such as vaccination, organ transplant, blood transfusion, and stem cell research, went through a similar initial opposition (Juma 2016).

In contrast to dichotomist categorizations, I have suggested that nootropics and other pharmacological neuroenhancers should be seen as part of a *neuroenhancement continuum*, together with several other cognition-enhancing substances and activities that include healthy nutrition, adequate sleep, education, information and computer technology, parental care, exercise, and many others.

After arguing that nootropics do not satisfy the conditions for *a priori* differential moral judgments compared to other enhancers, I have explored the epistemological grounds for guiding ethical decision-making with reference to nootropics and enhancers in general. In this context, I have called for an evidence-based approach to the evaluation of the risks and benefits of neuroenhancement. I represented this approach in the form of a simple decision-making heuristic, named MiRNA, in order to provide a simple and general tool for clinicians and other professionals to quickly assess the level of moral permissibility of any given enhancer. This heuristic could be advantageous for four main reasons: (a) it is applicable to all forms of enhancement, (b) it is quantitatively measurable and possibly modelizable, (c) it is computed from empirically testable variables, and (d) it might allow flexible, adaptive, and prompt decision-making.

Lastly, I applied MiRNA to a relatively wide range of nootropics currently in circulation to provide a rough index of current nootropics according to their degree of moral permissibility. This index consisted of three macro categories: (A) nootropics that are morally permissible without restrictions, (B) nootropics that are morally permissible with restrictions, and (C) nootropics that are impermissible (unless rare exceptions).

It is predictable that in the coming 20 years, many nootropics will be transferred from Category B to Category A, as advancements in neuropharmacology are gradually improving nootropics' safety and effectiveness. In managing this transfer, special attention should be reserved for the long-term effects on brain development, as there is no sufficient longitudinal evidence at the moment. In a world in which human life spans, work spans, work performances, and adaptive challenges are gradually increasing, nootropics and other neuroenhancers might become increasingly helpful for improving the quality of life, extending work productivity, as well as for mitigating normal and pathological forms of age-related cognitive decline. Safe and effective neuroenhancers will benefit not only the individual but society too. However, these benefits should be balanced with the associated risks.

References

Barkow, Jerome H., Leda Cosmides, and John Tooby. 1995. *The adapted mind: Evolutionary psychology and the generation of culture*. Oxford University Press.

Bonavita, Simona, R. Sacco, M. Della Corte, S. Esposito, M. Sparaco, A. d'Ambrosio, R. Docimo, A. Bisecco, L. Lavorgna, and D. Corbo. 2015. "Computer-aided cognitive rehabilitation improves cognitive performances and induces brain functional connectivity changes in relapsing remitting multiple sclerosis patients: an exploratory study." *Journal of neurology* 262(1):91-100.

Brayne, Carol, Paul G. Ince, Hannah A.D. Keage, Ian G. McKeith, Fiona E. Matthews, TuomoPolvikoski, and RaimoSulkava. 2010. "Education, the brain and dementia: neuroprotection or compensation?" *Brain* 133 (8):2210-2216.

Chee, Michael W.L., and Lisa Y.M. Chuah. 2008. "Functional neuroimaging insights into how sleep and sleep deprivation affect memory and cognition." *Current opinion in neurology* 21 (4):417-423.

Coveney, Catherine. 2011. "Cognitive enhancement? Exploring modafinil use in social context." In *Sociological reflections on the neurosciences*. Bingley: Emerald, 203-228.

Dees, Richard H. 2007. "Better brains, better selves? The ethics of neuroenhancements." *Kennedy Institute of Ethics Journal* 17 (4):371-395.

Feldman, Ruth, Zehava Rosenthal, and Arthur I Eidelman. 2014. "Maternal-preterm skin-to-skin contact enhances child physiologic organization and cognitive control across the first 10 years of life." *Biological Psychiatry* 75 (1):56-64.

Fontani, Giuliano, F. Corradeschi, A. Felici, F. Alfatti, S. Migliorini, and L. Lodi. 2005. "Cognitive and physiological effects of Omega-3 polyunsaturated fatty acid supplementation in healthy subjects." *European journal of clinical investigation* 35 (11):691-699.

Giurgea, Corneliu. 1973. "The 'nootropic' approach to the pharmacology of the integrative activity of the brain 1, 2." *Conditional reflex: a Pavlovian journal of research & therapy* 8 (2):108-115.

Greely, Henry, Barbara Sahakian, John Harris, Ronald C Kessler, Michael Gazzaniga, Philip Campbell, and Martha J Farah. 2008. "Towards responsible use of cognitive-enhancing drugs by the healthy." *Nature* 456 (7223):702-705.

Hillman, Charles H, Kirk I Erickson, and Arthur F Kramer. 2008. "Be smart, exercise your heart: exercise effects on brain and cognition." *Nature reviews neuroscience* 9 (1):58-65.

Juma, Calestous. 2016. *Innovation and Its Enemies: Why People Resist New Technologies*. Oxford University Press.

Leiva, Plaza B., Brito N. Inzunza, Torrejón H. Pérez, Gloor V. Castro, Medina J.M. Jansana, Díaz T. Toro, Flores A. Almagiá, Díaz A. Navarro, Cáceres M.S. Urrutia, and Oltremari J. Cervilla. 2001. "[The impact of malnutrition on brain development, intelligence and school work performance]." *Archivoslatinoamericanos de nutricion* 51 (1):64-71.

Lloyd, John Uri. 1911. *History of the Vegetable Drugs of the Pharmacopeia of the United States*. JU & CG Lloyd.

Maier, Larissa J., Matthias E. Liechti, Fiona Herzig, and Michael P. Schaub. 2013. "To Dope or Not to Dope: Neuroenhancement with Prescription Drugs and Drugs of Abuse among Swiss University Students." *PLOS ONE* 8 (11): e77967.

McCabe, Sean Esteban, John R. Knight, Christian J. Teter, and Henry Wechsler. 2005. "Non-medical use of prescription stimulants among US college students: Prevalence and correlates from a national survey." *Addiction* 100 (1):96-106.

Meyers, Chris D. 2014. "Neuroenhancement in Reflective Equilibrium: A Qualified Kantian Defence of Enhancing in Scholarship and Science." *Neuroethics* 7 (3):287-298.

Moreno, Jonathan D. 2012. *Mind wars: Brain science and the military in the twenty-first century*. Bellevue Literary Press.

Racine, Eric, and Cynthia Forlini. 2010. "Responding to requests from adult patients for neuroenhancements: guidance of the ethics, law and humanities committee." *Neurology* 74 (19):1555-1556.
Wu, Peggy, and C. Miller. 2005. "Results from a field study: The need for an emotional relationship between the elderly and their assistive technologies". In Edited by D. D. Schmorrow, *Foundations of Augmented Cognition*, Vol 11.

Blanca Rodríguez López

4 Procreative Beneficence: is Selection Really Better Than Genetic Modification?*

4.1 Introduction

At the beginning of the 21st century, Julian Savulescu proposed a principle intended to guide prospective parents' choices regarding the genetic makeup of their future children. The so-called Principle of Procreative Beneficence (PPB) states as follows:

> Couples (or single reproducers) should select the child, of the possible children they could have, who is expected to have the best life, or at least as good a life as the others, based on the relevant, available information. (Savulescu 2001, 415)

Some years later, together with Gay Kahane, he reformulated the PPB. The new formulation reads:

> If couples (or single reproducers) have decided to have a child, and selection is possible, then they have a significant moral reason to select the child, of the possible children they could have, whose life can be expected, in light of the relevant available information, to go best or at least not worse than any of the others. (Savulescu and Kahane 2009)

The new formulation clarified some points that were misunderstood in the original one (such as the meaning of "should") but did not change the PPB's essence. In both formulations, the PPB has been widely discussed, criticized by some and celebrated by others. PPB's impact is perfectly understandable, due to both the intrinsic interest of the issue and its own remarkable characteristics. PPB promotes an active and intentional role for the parents in the genetic makeup of their children, which is rejected by those who think that anything is wrong but letting nature have its way. PPB applies to what Savulescu calls "nondisease genes", thus entering into the enhancement territory, one of the most controversial in today's bioethics. And PPB focuses on the child's interests, and it speaks about the "best life", always a controversial concept. If all this were not enough, according to PPB, selecting a child is not merely a right the parents have, but something stronger. The possibility

* Derek Parfit died while I was working on the final version of this paper. This paper is humbly dedicated to his memory.

of selection gives the parents a "significant moral reason", putting them under some kind of moral obligation[1].

In this controversy between critics and supporters, I am rather in the second group. But though I generally agree with PPB, I would like to raise a different issue from the ones already mentioned, but unlike those, I do not consider it an essential part of the PPB. In this paper, after a few preliminary considerations, I examine PPB's focus on selection, which, for reasons explained later, seems very questionable.

4.2 Some considerations on the PPB

In this section, we consider some general points regarding PPB in order to facilitate the understanding of the principle before turning to the main object of this paper.

4.2.1 Definition of enhancement

In order to understand PPB's reference to the "best life", it is useful to say a couple of words about human enhancement. Defining human enhancement is not an easy task[2]. The usual strategy is to define "enhancement" by contrasting it with "treatment", as those interventions going beyond treatment. "In broad terms, therapy aims to fix something that has gone wrong, by curing specific diseases or injuries, while enhancement interventions aim to improve the state of an organism beyond its normal healthy state" (Bostrom and Roache 2008). Those unhappy with this definition because of its relation with the concept of health, problematic by its own right, try to talk about enhancement in terms not only of health but of human capacities. In contrast, the definition offered by Savulescu and Kahane is normative, as "any change in the biology or psychology of a person in a given set of social or environmental circumstances C that increases the chances of leading a good life in circumstances C" (Savulescu and Kahane 2009).

4.2.2 The best life

Genetic testing is usually performed in order to detect what Savulescu calls "disease genes" (2001, 415), genes that causes genetic disorders or predispose to suffer from some disease. One major reason is that, until recently, only tests for chromosome abnormalities such as Trisomy 21 (Down Syndrome) or single-gene disorders (such as

1 For a standard example of critics, refer Sparrow 2007. Refer also Birch 2005 and Stoller 2008.
2 For an excellent discussion of this problem, refer Juengst (1998).

cystic fibrosis) were available, and although some of them involve cognitive deficits, they mainly concern health. Another related reason is that until recently, the genetic contribution to non-health-related traits was not well known and even strongly denied by the predominant environmentalist view (Paul 1998; Pinker 2002). But all this is gradually changing: we know more about the genetic contribution to other traits and an increasing amount of tests are becoming available. We can reasonably expect that this increasing knowledge will contribute in the near future to undermine the predominant, strongly environmentalist view.

Apart from this, we can ask why we should, at both moral and prudential levels, care for health. As Savulescu states (2001, 417), the reason is no other than the relation between health and happiness, or health and welfare. And our welfare is much more than health. In our three main theories about well-being (Parfit 1976; Griffin 1986), health is far from being the only, or even the most important, thing. In the Objective List theory, it appears among the things that are part of our well-being; in the Hedonistic Theory, health only has an instrumental value because of the pleasures related with health and the pains caused by its absence in the Hedonistic Theory; finally in the Desire Fulfillment Theory, health plays a role because it enables us to do the things we want to do and to carry on the life we want to live.. And some of the things that contribute to our happiness depend, at least in part, on our biology. For instance, having a sunny disposition or the capacity for self-control contributes greatly to our well-being (Mischel 2015), and both seem to have a biological basis. Of course, there is no test available for many of these traits, and maybe there will never be for some of them, but, as far as we can, we have the same reason to test for them as we have to test for disease genes, and the PPB should apply to all of them.

For this reason, as Savulescu explicitly claims (2001, 2009), PPB advocates for selection not only regarding disease genes but also nondisease genes. If we admit that prospective parents have moral reasons to care not only about their future child's health but also about his/her potential for well-being, then couples should use genetic tests for nondisease traits, and selection should be allowed on this basis. For example, if the prospective parents have a choice of implanting one of two embryos that are genetically identical, except that one of them is genetically predisposed to higher intelligence, the parents-to-be are morally obliged to select the latter embryo over the other, since a more-intelligent child is likely to have a better life than a less-intelligent one, other things being equal. Now, we can turn to the aspect of PPB that I find less satisfactory because of its unduly narrowness.

4.3 Selection

There are four main ways for the prospective parents to make choices about their future children's genetic endowment: gamete selection, prenatal diagnosis and abortion, preimplantation genetic diagnosis (PGD), and manipulation of the genetic material of the embryo. The first three are cases of selection. In this section, I try to explore the reason why PPB focuses on selection[3].

4.3.1 Two (unconvincing) explicit reasons

In both versions of PPB, we can find two explicit reasons.

1. The practical reason

To begin with, there is a very sensible practical reason for favoring PGD. Once a couple[4] undergoes an in vitro fertilization (IVF) process, there is no added cost in performing PGD. Whatever the reason for IVF (fertility problems or fear of heritable genetic conditions), couples will probably be more inclined to test not only for severe genetic conditions but also for less-serious medical conditions or even for nondisease genes. After all, some embryos are going to be discarded anyway, and it is only good sense not to make the selection by tossing the dice.

However, not everything is rosy about selection via PGD. Some conditions can only by tested on fetuses. In these cases, selection means abortion. Abortion is costlier at all levels, and for many people, it is morally problematic.

It is also very costly to undergo an IVF process. Once you are having one, there is *no additional cost* involved for PGD, but to initiate one, you have to have some very good reason, as the process is very demanding at the physical, emotional, and psychological levels (Kaliarnta et al. 2011). This means that women are not likely to be willing to undergo IVF or to face an abortion only to have a child with better prospects of a better life, but only when under a high risk of having a child with really severe genetic conditions. And this is why this first reason is unconvincing: it is only effectual for a very small portion of the population, those already having IVF for another independent reason.

2. The continuity reason

Another reason to focus on selection is continuity with generally admitted practices. We traditionally select a child when we select a partner or when we choose the moment of conception, not only for health related reasons but also for economic and social ones (Savulescu and Kahane 2009, 276): people who want to have a child usually try to choose a moment with favorable conditions, for instance, when you have economic security, or a house, or a good job, so that you can provide the child

3 The focus is on selection via PGD or prenatal testing. In the first version of 2001, the focus was even narrower, limited to PGD, and even in 2009, though mentioning prenatal screening, the main focus remains on PGD.

4 I completely agree with Savulescu's qualification "couples or single reproducers", but I will use "couples" for the sake of shortness.

with better conditions. And when someone has a child in suboptimal conditions, at least when these can be reasonably expected to change for the better in the near future, we think this to be morally wrong. Of course, people are not usually aware of being selective when they do all these things, but it is easy to show that they in fact are. You are here because you were born from a particular ovum and sperm. If your mother would have waited 1 year to have a child, or even just 1 month, another child would have been born, from another ovum and sperm.

This is undoubtedly true, but in order to convince people to follow PPB, it is not enough to show that to select our children is an accepted practice. The methods used to select are far from immaterial, and some of them are controversial. PGD is morally problematic for some people, and prenatal diagnosis followed by abortion is morally problematic for many. As often happens in morals, means matter.

4.3.2 One (unconvincing) implicit reason?

There is one reason that we can refer to as "state-of-the-art reason". Selection can be performed *now*. IVF plus PDG is readily available, whereas manipulation of the genetic material of embryos seems to be almost science fiction.

There is some truth in this. Until recently, genetic testing had been the only way to control a variety of genetic traits and conditions: chromosomal abnormalities, such as trisomy 21 (Down syndrome), single-gene disorders such as cystic fibrosis, some inherited cancer syndromes, adult-onset neurological conditions such as Huntington and Alzheimer's diseases, sex, or even minor disabilities such as deafness.

However, the possibility of making genetic choices not involving selection probably is not so far away in the future. In the past few years, many advances have been made in gene therapy and, though its development is extremely difficult[5], we are not unreasonably optimistic if we expect some degree of success in the near future. This optimistic view has recently gained some support since, in 2012 (Gyngell 2017), an influential paper was published proposing the clustered regularly interspaced short palindromic repeats (CRISPR)/CRISPR-associated proteins (Cas9) technology as a tool for gene editing, starting what is now known as the CRISPR revolution (Jinek et al. 2012).

Even if some conditions remain elusive to this kind of therapy, for others, we are close to success. For instance, this is the case with hemophilia. This is a heritable genetic condition that can be tested on carriers and also prenatally. If a gene therapy

5 The condition targeted must be well understood, the underlying faulty gene must be identified and a working copy of the gene must be available, the specific cells in the body requiring treatment must be identified and accessible, and the means for delivering working copies of the gene to these cells must also be available.

were available, to perform it would not involve selection. It would be prenatal genetic manipulation.

4.3.3 The real (explicit) reason

There is a difference between the various ways of affecting your future children's genome mentioned herein. The ones that we labeled "selection" (gamete selection, prenatal diagnosis and abortion, and PGD) are identity affecting, whereas the manipulation of the genetic material of the embryo can be considered a non–identity-affecting route. When prospective parents choose to implant one particular embryo after IVF, to have an abortion after screening and wait for the next pregnancy, or to sort sperm for fertilization, the child they are going to have as a result is a different child from the one they would have had had they made a different choice. The same can be said if they decide to postpone pregnancy until they have achieved a good economic situation. This is the reason why in these cases, we talk of selection. On the contrary, if you perform some genetic manipulation in your embryo or fetus, the child is the same one but, let us say, without propensity to suffer from asthma. The result is the same as if you have performed some gene therapy for preventing asthma in your 3-year-old child. If you treat your asthmatic child with this kind of therapy, you are not replacing him/her with a different one[6]. If your choice is affecting the identity of a (future) child, then you may face the nonidentity problem.

In Part four of his famous and very influential book, entitled "*Future Generations*", Parfit (1986) addresses some moral questions related to those of our actions that can affect people who do not yet exist. Apart from its influence on the question about the structure of our moral theories (basically whether they are to be person-affecting or impersonal), they have had a remarkable influence on bioethics. After devoting the third part of the book to the very intricate question of personal identity over time, he then asks "what would have made it true that some particular person would never have existed?" We exist, but we could have not existed. Some of the decisions we make (and not only what we properly call "procreative decisions") affect the identity of future people (and not only their number). This gives raise to the nonidentity problem that Parfit illustrates with some examples. The first one is the relevant one for us here. Imagine a teenager who decides to have a child. Due to her youth, her

6 Of course, in a sense, the resulting nonasthmatic child would be different from the asthmatic one, as far as our identity is built from our experiences and the kind of life we live; but the same can be said if the therapy is not a genetic one, and in both cases, the difference between "the child before" and "the child after" would not be greater than the one between the child's situations if you send him/her to secondary school or not. In this biographical sense, every parental choice influences the identity of the child, but these choices are identity-preserving ones in a relevant sense.

child will probably have a bad start in life. If she waits, she will have a child with better prospects. But it will be a different child. Should she wait? If you think so (as almost everyone does), and try to give a reason why, you will find that this reason cannot be related with any harm suffered by someone. In making this decision, she is *selecting* which child to have. If she chooses to have a child now, *this child* (A) has no reason to complain, as far as his/her life is worth living, because the only way of giving him/her the best start in life is waiting and, as a consequence, not having *him(her)*, but some other child (B). So, you can say: you exist only because you had a bad start in life, so if you are happy enough to exist, you have nothing to complain about.

Savulescu thinks that this problem is the main reason to prefer selection over genetic manipulation (Savulescu et al. 2006). So, we can safely assume that, in his formulation of PPB, he focuses on selection for this reason. Savulescu claims that selection is to be favored over genetic manipulation, not only because of the practical and sensible reasons commented earlier, but also, and mainly, because it is less morally problematic. As this claim is far from obvious, I will address it in the following section.

4.4 PPB and the nonidentity problem

Savulescu's real reason to focus on selection is that it is less problematic, at a moral level, than genetic intervention. To hold that selecting is less morally problematic requires some further assumptions.

The first one refers to the possible harm to the future child (Savulescu 2001). In any choice you make, there is always the possibility of something going wrong. When you make a choice regarding the genome of your future children, with good intention of increasing their chances of leading a good life, there will always be a risk of reduced well-being. If something goes wrong, and your choice has involved selection, then the child has not been harmed, for the simple reason that if he/she would not have been selected, he/she would not have existed. He/she has not been harmed as far as his/her life is worth living. On the other hand, if your choice has involved genetic manipulation, the child has been harmed since he/she could have existed in a better condition.

The second one relates to the possibility of your child having, in the future, a fair ground for complaining, even if nothing has gone wrong. For different reasons that would take us too long to analyze here, there is a possibility of choosing a genetic trait that, later in life, your child would prefer not to have had. As in the previous case, and for the same reasons, if your choice involved selection, he/she could not reasonably complain about having it. But in the case of genetic intervention, the resulting child may complain, because without this intervention, he/she could have existed (he/she would still have been himself/herself) and have what he/she could consider a better life.

In short, as far as your child has been selected, and his/her life is worth living, there is no way for you to have done him/her harm, since the very condition for him/her to exist is to be the way he/she is. But if you have changed your child's genome, then you could have harmed him/her.

Harming someone is morally wrong in any sensible view, though not necessarily the only way of doing something morally wrong. In fact, Savulescu claims that, by not selecting the child with a better prospect of living a good life, you do something morally wrong. This is implied by saying that you have a moral reason to select such a child. But even if you do not hold (as Savulescu does not) a person-affecting view of morality (Parfit 1986), selection seems morally superior to genetic intervention as long as you hold the less-radical (and probably more reasonable) view that, though impersonal considerations also have a place in morality (i.e., that some harm can be done even if nobody is harmed, only because the world would be a worse place because of containing less happiness), nonetheless, these impersonal considerations matter less than personal ones.

I share this view of morality. I do think that impersonal considerations matter and that personal ones matter more. Nevertheless, I think this reason to favor selection over genetic enhancement is not convincing. I will explain why in the following section.

4.5 Why selection is not morally superior

My first, and less important, reason is that the previous arguments are too dependent on a controversial view of identity. For some, holding what DeGrazia calls the thesis of fragile prenatal identity (DeGrazia 2005), prenatal genetic manipulation can also be identity affecting (Zohar 1991). Though I do not support this view, it is worth keeping in mind that there is no universal agreement on this question.

The second reason relates to the alleged moral wrongness of harming someone. If I harm you by putting eggs in your breakfast, in ignorance of you being allergic to them, I do nothing morally wrong. If I send my daughter to horse-riding training courses, and if she falls and breaks her leg, or meets someone there who latter on has a bad influence on her life, I do nothing morally wrong. And she cannot fairly complain. Many prenatal, perinatal, and postnatal activities are risky for the child. Some (most) of them are taken for the sake of the child, such as medical procedures and leisure or educational activities. On some versions of consequentialism, we have moral reasons to take into account expected consequences. When our choices involve risk or uncertainty (i.e., virtually always), expected consequences and expected well-being is all we have.

Of course, the possibility of harming our child (and of doing something morally wrong and of our child having fair grounds for complaining) puts some limits on what you can do (genetically or otherwise) to your child. You need good reasons to think that something (a course, a medical procedure, or a diet) is going to be good for him/her. You need good reasons to think that to give him/her certain genetic traits is going to be good for him/her, that is, it is going to give him/her the best (or as good as) probability of living the best (or as good as) possible life. Probably you also have to make sure that your idea of a good life is not too controversial or parochial, so that your child could share it in the future, or that the genetic choices you make for him/her (or his/her training, diet, and so on) are not going to put the child in a very narrow path (that is, he/she will have a wide set of alternatives to choose his/her own life in the future and develop and exercise his/her autonomy) amid all the caveats you wish to make. But this is a *good thing*. In fact, I consider that one of the most unpalatable consequences of selection, and of the reasons offered in support of its alleged superiority, is that it lets parents too easily off the hook: they can do a lot of otherwise-morally-questionable things and, after all, they are not harming anyone; the child cannot complain. It is a good thing because it would properly stress the fact that in making these kinds of choices, we are in the arena of parental autonomy, where we have a certain consensus about the limits: children are not parents' property and they cannot do to them whatever they fancy.

My third reason is that genetic intervention, unlike selection, does not sound eugenic. This is a very important point. Any time you talk about PPB, the first question asked is always the same: is this not eugenics? This question is not unreasonable. Leaving apart the various and widely discussed, undesirable and seriously morally unacceptable characteristics of old eugenics (compulsory, racist, state directed, and so on), its goal is considered by many as, to say the least, highly problematic, as far as it is intended neither to cure nor to enhance people and people's lives, but to replace them. Selection means to decide to bring some people into life instead of others (if you select by PGD) and, for many people, if selection is performed via abortion, it involves eliminating one person and replacing him/her with another one. I think this is the reason why although some people accept selection to avoid some devastating genetic conditions, they strongly oppose selection when the condition is not so extreme, not to mention the possibility of selecting, as the PPB asks, for nondisease genes.

But genetic intervention does not, in any sensible view, eliminate people. Because it is not identity affecting, it *changes people*. In this respect, it is similar to therapy, and we usually accept therapy for not very critical conditions, and addressing nondisease genes is not dissimilar to schooling, training, and other widely accepted child-rearing practices. Of course, some people can still oppose genetic interventions but for reasons unrelated to the eugenic complaint.

4.6 Conclusion: a defense for an extension of PPB

PPB focuses on cases of selection. But it is silent on genetic interventions. Many philosophers seem to think that selection is less morally problematic[7]. I am not sure at all whether this view can be easily accepted by nonphilosophers. I know better than to generalize from my personal experience, but I always find it easier to convince people about gene therapy than about selection, even when talking about enhancement. But maybe this is because I probably argue for it more persuasively. In this paper, I have tried to explain my reasons, and now I only want to stress a couple of issues.

There can be some reasons to choose selection over genetic intervention or the other way around, and probably different people have different reasons. But the moral superiority of selection is questionable. In fact, it is more than questionable, at least if we focus on the children's well-being, as PPB does. The nonidentity problem can be easily used to make controversial choices about future people, including our own children. You can discharge your responsibility saying "if you were not deaf, have achondroplasia, be without clitoris, suffer from depression and so on, you would simply not exist at all"[8]. And this seems quite dangerous to me, and contrary to what I think is the very spirit of PPB. If you think, as I do, that PPB is a reasonable principle, and you consequently think controversial choices should not be made for the children[9], I think the nonidentity problem does not provide a reason to consider selection morally superior.

There seems to be no good reason not to extend PPB to cover those cases that do not involve selection but genetic intervention. And doing so has some advantages. Firstly, this method would be less costly than selection both via abortion and via PGD. It would be less painful both physically and psychologically for the parents and would also be less morally problematic, since it does not involve abortion or embryo destruction. Secondly, we could also show continuity with generally admitted practices, such as giving vaccines to your child, sending him/her to school, or caring about his/her diet or physical activity. Thirdly, it does not sound eugenic, or is at least less eugenic. While eugenics tried to eliminate some people or not to let some people to be born, genetic intervention, being identity preserving, aims to change people for the better.

If my arguments are sound, we should extend PPB, changing its formulation accordingly to include not only selection but also genetic intervention. It would read as follows:

> "If couples (or single reproducers) have decided to have a child, and it is possible to choose which kind of child to have, then they have a significant moral reason to choose the genetic endowment that, in light of the relevant available information, gives the child the best chance of having the best possible life."

---

7 Among the many, refer Smolensky 2008.

8 For a more detailed account, refer Smolensky 2008 and Cohen 2009.

9 It is an open question whether some other moral considerations can override PPB, among them – and prominently – the well-being of parents.

References

Birch, Kean. 2005. "Beneficence, Determinism and Justice: An Engagement with the Argument for the Genetic Selection of Intelligence." *Bioethics* 19 (1): 12–28.

Cohen, I. Glenn. 2009. "Intentional Diminishment, the Non-Identity Problem, and Legal Liability". Hastings Law Journal, Vol. 60, 2008; Harvard Public Law Working Paper No. 09-11. Available at SSRN: https://ssrn.com/abstract=1330504

DeGrazia, David. 2005. *Human Identity and Bioethics*. Cambridge University Press.

Jinek, Martin et al. 2012. "A programmable dual-RNA-guided DNA endonuclease in adaptive bacterial immunity." *Science* 337: 816-21.

Feinberg, Joel. 1980. "A Child's Right to an Open Future." In *Whose Child? Parental Rights, Parental Authority and State Power*, ed. W. Aiken and H. LaFollette, H., Totowa, NJ: Littlefield, Adams, and Co., 124–153.

Griffin, James. 1986. *Well-Being. Its Meaning, Measurement, and Moral Importance*. Oxford. Clarendon Press.

Gyngell, Christopher. 2017. "Gene editing and the health of future generations." *Journal of the Royal Society of Medicine* 110 (7): 276-279.

Kahane, Guy, and Julian Savulescu. 2009. "The Welfarist Account of Disability." In A. Cureton and K. Brownlee, eds., *Disability and Disadvantage*. Oxford University Press, 14-53.

Kaliarnta, Sofia, Jessica Nihlén-Fahlquist, and Sabine Roeser. 2011. "Emotions and Ethical Considerations of Women Undergoing IVF-Treatments", *HEC Forum* 23:281–293

Mischel, Walter. 2015. *The Marshmallow Test: Understanding Self-control and How To Master It*. Penguin

Paul, Diane B. 1998. *The Politics of Heredity*. State University of New York Press.

Parfit, Derek. 1986. *Reasons and Persons*. Oxford University Press.

Pinker, Steven. 2003. *The Blank Slate*. New York, NY: Viking.

Robertson, John A. 2003. "Procreative liberty in the era of genomics." *American Journal of Law and Medicine*, 29: 439-87

Savulescu, Julian. 2001. "Procreative Beneficence: why we should select the best Children." *Bioethics* 15(5/6): 413-426.

Savulescu, Julian. 2007. "Genetic Interventions and the Ethics of Enhancement of Human Beings." In Steinbock, B. (ed). *The Oxford Handbook of Bioethics*. Oxford: Oxford University Press, 516 – 535.

Savulescu, Julian. 2009. "Genetic Enhancement." In H. Kuhse and P. Singer (eds). *A Companion to Bioethics: 2º edition*. Wiley-Blackwell, Oxford, 216-234.

Savulescu, Julian, and Kahane G. 2009. "The moral obligation to create children with the best chance of the best life." *Bioethics* 23(5): 274-290.

Savulescu, Julian, et al. 2006. "Behavioural Genetics: Why Eugenic Selection is Preferable to Enhancement." *Journal of Applied Philosophy* 23(2): 157-171.

Singer, Peter. 1993. *Practical Ethics*, 2º edition. Cambridge University Press.

Singer, Peter. 2003. "Shopping at the Genetic Supermarket." In S. Y. Song, Y. M. Koo and D. R. J. Macer (eds.) *Asian Bioethics in the 21st Century*, Tsukuba, 143-156.

Smolensky, Kirsten Rabe. 2008. *Creating Children with Disabilities: Parental Tort Liability for Preimplantation Genetic Interventions*, 60 Hastings L.J. 299, 336.

Sparrow, Robert. 2007. "Procreative Beneficence, Obligation, and Eugenics." *Genomics, Society and Policy* 3(3): 43-69.

Stoller, Sarah E. 2008. "Why We Are Not Morally Required to Select the Best Children: A Response to Savulescu." *Bioethics* 22 (7): 364–9.

Macer, Daryl R. J. 2003. *Asian Bioethics in the 21st Century*. Tsukuba, 143-156.

Zohar, Noam J. 1991. “Prospects for ‘Genetic Therapy’ – Can a Person Benefit from Being Altered?” *Bioethics* 5: 275-288.

Daniela Rusnac

5 Practical Ethics Issues in Gene Therapy and Genetic Testing

5.1 Introduction

The development of biotechnologies, as well as the emergence of genetic engineering and other innovations, promises great benefits, but it could also have a negative impact on the development of the human species and nature. Genetics is an area of research with an impressive history of evolution and significant practical advances that influence human experience. In 1953, Crick and Watson identified the structure of DNA. By the early 1970s, scientists succeeded in isolating genetic material from one species and attaching it to the genetic material of other species, thereby bringing about genetic engineering. In the same decade, the foundation for a program in gene therapy was established via the discovered possibility of identifying the genetic background of different pathologies.

During its development, this field of research gathered supporters and opponents. Supporters point to successful achievements, such as solving the food problem, creating food with a higher medical value, potential development of food vaccines, treatment of genetic mutations, prevention of diseases in children in the prenatal period, extending the human life span, and prevention of diseases using genetic tests. Opponents highlight the risks of genetic defects, limited genetic diversity, increasing inequalities, and emergence of eugenics. This article analyzes some ethical issues that arise in practice from the use of genetic testing, gene therapy, and germ-line therapy, topics that are increasingly discussed in European context.

5.2 Genetic testing

By the early 1980s, scientists envisioned the possibility of mapping all of the human genes, thereby laying the foundation for the Human Genome Project. Genetic testing can be used to precisely identify a disease or to find the faulty gene that increases the probability of a particular disease's onset in a person. Nowadays, it is possible to predict the development of a large range of disorders, such as Huntington's chorea, Alzheimer's disease, muscular dystrophy, hemophilia, multiple sclerosis, and different types of cancer. Genetic testing is used to detect carriers of faulty genes who may show no signs of the disease but can pass it on to their children. For instance, if both parents are carriers of cystic fibrosis, then it passes almost with certainty to their child. Genetic testing is used to identify a fetus with certain diseases such as Down syndrome, as well as to screen newborns for the same disease. For instance, newborn screening is widely practiced in the case of the hereditary metabolic disorder phenylketonuria (PKU). If the blood test is positive for PKU, a special diet is used to avoid the buildup of an enzyme that causes brain damage (Buchanan et al. 2000, 13).

Despite its therapeutic and preventive advantages, application of genetic testing raises ethical issues in practice, mostly related to technological limitations of genetic testing, ownership of genetic information, and potential for discrimination. The tests cannot always identify the mutation, and even if they did, a faulty gene is not necessarily indicative of future symptoms, such as the severity of the condition or age of onset. Unawareness about the limits of genetic testing can set false expectations and lead to dissatisfaction with medical service or to unmanaged psychological distress.

But widespread application of this new technology faces many social issues. Who should be tested? Why? How much does it cost and who should pay for it? Currently, most doctors accept that only those with a familial risk should be tested for a specific disease. Because this technology is very expensive, it prohibits people from having unnecessary screening and limits access for low-income and marginalized groups.

Further, should parents have the right to choose the traits of their children beyond therapeutic goals? Would it be right for a community to create designer babies because they decide that it is their right to choose what they want, and what would be best for their child? One way to approach this issue is to claim that prenatal genetic screening is morally permissible as long as the risks to the mother and fetus do not outweigh the benefits of the information gained by the test. This kind of testing should be allowed if it is undertaken for the purpose of early intervention.

Another issue is devising effective means of protecting privacy (confidentiality) and developing criteria for a voluntary screening program. Among these criteria are the following: a) the program should involve tests that are highly sensitive and specific and that have high positive predictive value; b) there must be available therapy or other interventions that are more useful if applied before symptomatic disease appears or the knowledge gained must be otherwise valuable to the individual being screened; c) the program can be justified economically in comparison to other ways in which health funds could be used; d) the program must be based on laboratory work whose high quality can be assured; and e) the program must involve adequate counseling before the screening to ensure informed consent as well as adequate counseling about the meaning of the results afterward (Brody 1998, 90-91). According to the principle of confidentiality, access to patient outcomes by a third party is prohibited without the express consent of the patient.

Ownership of genetic information is extremely difficult to determine. On the one hand, information belongs to whoever underwent the test, but on the other hand, genetic information is about family and consequently is relevant beyond the tested individual. A physician thus faces the moral dilemma of how to balance individual privacy and the responsibility of preventing harm to other family members. The decision to inform family members vary according to what moral theory is used (Fulda and Lykens 2006). According to the utilitarian theory, the decision to communicate genetic test results to a third person should be based on a cost–benefit analysis. The benefits for the family must prevail over the benefits of the tested person because the moral principle of the utilitarian theory is to ensure maximum happiness. Not only the family, but also other members of the society, will be able to minimize the risk of developing a disease in the future. Thus, an important task is the assessment of risks and benefits of disclosure according to the complexity of the situation. On the other hand, the most important value of libertarians is moral autonomy: informed decision-making is a vital right of each person. Individuals also have the right to confidentiality, or to choose whether they want to inform their family members. Tested subjects have the right to keep the medical results secret because, over time, they can develop a fear of being discriminated against, if relatives find out that in the future they will develop a certain disease. From a duty-based perspective, we have to respect the command of beneficence for society (or family) and to disclose the result of the test, but at the same time, a physician could break the duty of nonmaleficence for the tested person. Thus, for each case, we have to take into account a risk–benefit ratio for the tested person.

Genetic information has a high potential for discriminating usage, such as discrimination in insurance, discrimination in employment, racial discrimination, and determination of paternity. Knowing that the purpose of a company is making profit, it is plausible to assume that companies that have access to genetic information would be incentivized to prefer healthy employees rather than potentially ill employees. Genetic testing could be used in an inappropriate manner for determining paternity without the informed consent of all parties involved or solely for deciding whether to terminate pregnancy because of gender. Take, for instance, the chilling discrepancy in sex ratios in China, South Korea, and parts of India, where boys outnumber girls by up to 30% (Eberstadt 2002). For example, in 2001, in China were missing 34–41 million females, in India 27–39 millions, and in South Korea 0.2–0.3 million females (Hesketh 2006).

Thus, genetic screening has many benefits, but it also has downsides when we consider its application in practice. To list just a few benefits: early detection of any type of disease or disorder is usually less expensive to treat; genetic screening helps people know whether a heritable disease runs in their family; paternity testing is helpful for single mothers. To list the highlighted downsides: health coverage is limited because of high costs; ensuring confidentiality is difficult; and genetic screening could increase abortion rates.

We also could add that voluntary choice should be the basis for genetic testing. For example, The World Medical Association Declaration on the Human Genome Project declares that “One should respect the will of persons screened and their right to decide about their participation and about the use of the information obtained”.

Thus, to ensure the genetic health of the population, the European Commission has established recommendations on genetic testing to explore the conditions for maximizing benefits and minimizing harms from the increasing capability to use genetic testing. Here are the recommendations that address the ethical issues discussed so far:

R.9 – In the context of healthcare, genetic testing be accompanied by the provision of key information and, where appropriate, by the offer of individualised counselling and medical advice (in the case of highly predictive genetic tests for serious disorders, the offer of specific counselling should be mandatory, and patients should be strongly encouraged to take advantage of it).

R.10 – Genetic data of importance in a clinical and/or family context should receive the same level of protection as other comparably sensitive medical data; b. the relevance for other family members has to be addressed;

R.11 – Data derived from genetic sources should not be used in ways that disadvantage or discriminate unfairly against individuals, families or groups in either clinical or non-clinical contexts, including employment, insurance, access to social integration, and opportunities for general well-being;

C – Timely access to genetic testing should be based on need and appropriately resourced with no discrimination based on gender, ethnic origin, social or economic status. (United Nations Educational, Scientific and Cultural Organization [UNESCO] 2004)

Accordingly, the European Commission suggests that the responsibility of dissemination of genetic test results should be assigned to the family, highlighting the necessity of detailed discussion both prior to testing and even after it, as well as recommending the avoidance of discriminatory use of genetic data. However, further oversight should be put in place to see how these recommendations are respected in practice.

One output of the human genome project is the accurate identification of individuals who will develop various diseases requiring very expensive treatment. Thus, we must remain vigilant to detect any potential patterns of discrimination against those who are genetically predisposed to various illnesses. The goal of health coverage is to spread risks among large populations regardless of genetic heritage. Similarly, equal opportunity for employment mandates that genetic heritage should not become the basis for employment selection. Even properly validated genetic testing should never be mandated for insurance or employment proposes, and those who are voluntarily tested should have total control over who receives the results. Thus, according to the insurance principle adopted by the Council of Europe, "insurers should not have the right to require genetic testing or to enquire about results of previously performed tests, as a precondition for the conclusion or modification of an insurance contract" (Council of Europe 1992). It is increasingly clear that the principles of autonomy, confidentiality, and nondiscrimination are put to the task of new applications and interpretations in the context of genetic testing, release of genetic information, and use of genetic information in employment and insurance decisions.

5.3 Gene therapy

Gene therapy aims to treat incurable illnesses – resulting from defective genes – by the insertion of healthy genes into an individual's affected cells and tissues. The technique is tested on viruses because viruses are good at injecting their DNA payload into human cells and reproducing it.

In the initial stages of treating genetic disorders, scientists were confident that gene therapy would curb a vast amount of human suffering due to painful, and sometimes fatal, genetic diseases. But since 1980, when the first experiment using these techniques was conducted, the initial enthusiasm has become tempered, mostly because of increasing side effects. Important media outlets such as the *New York Times* wrote about these cases: "The Arizona patient, Jesse Gelsinger, 18, died four days after doctors at the University of Pennsylvania injected a corrective gene, encased in a deactivated adenovirus, into the hepatic artery, which leads to the liver (...) Leading scientists and government officials said Gelsinger was the only person known to have died as a direct result of receiving gene therapy. Other patients receiving gene therapy have died, but doctors treating them in the trials say their diseases, not the therapy, killed them. What remains unknown is how many patients experienced side effects directly related to gene therapy and what they were" (Stolberg 1999). The lack of expected results leads to the acknowledgment that much more basic research is required, but lack of success did not burn out the optimism for gene therapy.

Genetic engineering, which sometimes is called genetic modification, is the process of altering the DNA in an organism's genome. Usually genetic engineering is used by scientists to enhance or modify the characteristics of an individual organism. Genetically engineered bacteria and other microorganisms are currently used to produce human insulin, human growth hormone, a protein used in blood clotting, and other pharmaceuticals, and this number will only increase in the future.

Enhancing cognitive capacities beyond human limits is still a promise for the future, but the basic argument in favor of doing it is so that it could make life better by improving intelligence, beauty, endurance, certain personality characteristics, and behavioral tendencies. If these traits were found to have a genetic basis, we could enhance people by intervening in the genetic makeup. However, such a technique could be dangerous in discriminating against persons with disabilities. If we lack a robust system of distributive justice for bioenhancement services, then discrimination for those without access to these services is expected (Beauchamp 2013). People with disabilities are often discriminated against by having fewer opportunities than others. By removing genetic disorders, and the resulting impairment, it is true that gene therapy could remove this source of inequality, but an implicit assumption is that people with genetic disorders need to be treated and made "normal". The objection sees gene therapy as a form of discrimination against impaired and disabled people.

The last point is that human enhancement could lead to a resurgence of eugenics in new forms. In the past, eugenics was used to justify practices including involuntary sterilization and euthanasia. For example, "the eugenic movement in the United States in the early 1900s was based on the use of 'genetic science', which claimed that complex social traits such as alcoholism, criminality, and depression were inherited in a simple fashion. This 'science' was cited as justification for policies that restricted immigration into the U. S., as well as for state law that permitted involuntary sterilization of criminals, mentally disabled individuals, and even unwed mothers. However, eugenics research was ultimately dismissed as significantly flawed, and the movement ceased by 1940" (Norrgard 2008). However, many now defend a new brand of eugenics by which individuals are free to choose whether to use genetic technologies for reproductive purposes (Buchanan et al. 2000; Agar 2004). One has to ask whether reproductive liberty is the absolute value for guiding policy in this context.

Another issue that has sparked the debate is what conditions should be treated by gene therapy. Proposed policies set two limits for gene therapy: a) gene therapy should not be pursued for nontherapeutic aims; b) gene therapy should not involve germ-line therapy.

Proponents for germ-line gene therapy argue that it is morally wrong to allow children to be born with fatal genetic diseases when the capability exists to remove those genes from the population once and for all (Savulescu et al. 2015). They also claim that this kind of therapy is a significant tool to curb human suffering, as well as a practical measure against the high costs of conventional treatment for generations of people afflicted with a given disease.

By contrast, opponents argue that it is inadvisable at this time because so little is known about gene regulation or the mechanisms of embryological development. They claim that the premature use of such techniques could be even more harmful than the disease they are trying to cure. It is acceptable to change the genome of a mature individual after medical indications or on his/her own desire, from an ethical point of view. But "a different situation occurs when changing the genes of embryonic cells because

1. this action can be described as amoral, due to the fact that the given research is carried out on unborn individuals;
2. an experiment on the human genome resulting in failure cannot be corrected;
3. an improperly constructed genome can be spread (through heredity);
4. the interaction character of 'new' genes with their totality is not yet fully studied, and reorganization of the genome of the embryonic cells can lead to unpredictable consequences" (Melnov 2016, 218).

An additional concern is that not only will diseases be selected out from the population, but also relatively insignificant problems such as myopia, racial variations such as skin color, and normal variations such as height. Opponents generally argue that germ-line therapy is a slippery slope that will push humanity into eugenic practices, thus resulting in a reduction of diversity in the human gene pool, which could increase our collective susceptibility to new diseases.

The International Summit on Human Gene Editing, which took place in Washington in 2015, reached several conclusions about how gene editing should be handled.

5.3.1 Basic and preclinical research

Intensive basic and preclinical research is clearly needed and should proceed, subject to appropriate legal and ethical rules and oversight, on (i) technologies for editing genetic sequences in human cells, (ii) taking into account the potential benefits and risks of proposed clinical uses, and (iii) understanding the biology of human embryos and germ-line cells.

5.3.2 Clinical use: somatic

Many promising and valuable clinical applications of gene editing are directed at altering genetic sequences only in somatic cells – namely, cells whose genomes are not transmitted to the next generation. Because proposed clinical uses are intended to affect only the individual who receives them, they can be appropriately and rigorously evaluated within existing and evolving regulatory frameworks for gene therapy, and regulators can weigh the risks and potential benefits in approving clinical trials and therapies.

5.3.3 Clinical use: germ line

Gene editing might also be used, in principle, to make genetic alterations in gametes or embryos, which will be carried by all of the cells of a resulting child and will be passed on to subsequent generations as part of the human gene pool. Although each country has its legal framework in different areas of biomedical research, in terms of the human genome, it must be regulated similarly in all countries.

It would be irresponsible to proceed with any clinical use germ-line editing unless (i) the relevant safety and efficacy issues have been resolved, based on appropriate understanding and balancing of the risks, potential benefits, and alternatives, and (ii) there is a broad societal consensus about the appropriateness of the proposed application. Moreover, any clinical use must proceed only under appropriate regulatory oversight. At present, these criteria have not been met for any proposed clinical use: the safety issues have not yet been adequately explored; the cases of most compelling benefit are limited; and many nations have legislative and regulatory bans on germ-line modification. However, as scientific knowledge advances and societal views evolve, the clinical use of germ-line editing should be revisited on a regular basis (International Summit on Human Gene Editing 2015).

Scenarios on the desired futures for human gene editing differ widely. Several participants in the discussion agreed with the idea that no new biomedical technology is safe. Even though the level of acceptable risk is subject to intense disagreement, gene editing will become acceptable when its benefits, both to individuals and to the broader society, exceed its risk. Some have proposed a moratorium on the basic research needed to enable germ-line human gene editing until an international ban on germ-line gene editing for reproductive purposes can be secured through the United Nations and all countries have adequate regulations for such research (International Summit on Human Gene Editing: A Global Discussions 2015).

5.4 Conclusion

In order to protect humanity from the potential danger that may occur from the implementation of emerging technologies such as genetic engineering, both scientists and policymakers have sought some safe and effective models. This objective could be achieved only after the concept of the precautionary principle has been elaborated. This principle enables decision-makers to adopt precautionary measures when scientific evidence about the environment or human health is uncertain and the stakes are high. Thus, the precautionary principle is understood "as the rule that one should never engage in the technological development or application unless it can be shown that this will not lead to large-scale disaster or catastrophes" (Engelhard & Jotterand 2004, 303). Nevertheless, we think that scientists should have the responsibility to reflect on the potential applications of their research against a plurality of values, for minimizing ethical complications in genetics research.[10]

References

Agar, Nicholas. 2004. *Liberal eugenics: In defence of human enhancement.* Oxford, UK: Blackwell.
Beauchamp, Tom L. 2013. "Are we unfit for the future?" *Journal of medical ethics*: medethics-2013.

10 I would like to thank Adriana Paladi and Emilian Mihailov for their guidance and comments in writing this paper.

Buchanan, Allen, Dan W. Brock, Normal Daniels, and Daniel Wikler. 2000. *From Chance to Choice: Genetics and Justice*. Cambridge: Cambridge University Press.

Council of Europe. 1992. Recommendation No. R(92)3 of the Committee of Ministers to member states on genetic testing and screening for health care purposes. *J Int Bioethique* 3:255–257.

Council of Europe. 1992. Recommendation No. R(92) 3 of the Committee of Ministers to Member States on genetic Testing and Screening for Health care Purposes. Available at https://wcd.coe.int/com.instranet.InstraServlet?command=com.instranet.CmdBlobGet&InstranetImage=573883&SecMode=1&DocId=601492&Usage=2

Eberstadt, Nicholas. 2002. Testimony to President's Council on Bioethics, October 17, at http://www.bioethics.gov/transcripts/oct02/session2.html

Engelhard, Tristram H., and Jotterand Fabrice. 2004. "The Precautionary Principle: A Dialectical Reconsideration". *Journal of Medicine and Philosophy*, 29(3), 301-312.

Fulda, Kimberly G., and Kristine Lykens. 2006. Ethical issues in predictive genetic testing: a public health perspective. *Journal of Medical Ethics*, 32(3), 143–147.

Hesketh, Therese, and Wei Xing Zhu. 2006. "Abnormal Sex Ratios in Human Populations: Causes and Consequences." *Proceedings of the National Academy of Sciences of the United States of America* 103(36), 13271–13275.

International Summit on Human Gene Editing. 2015. Available at http://www8.nationalacademies.org/onpinews/newsitem.aspx?RecordID=12032015a (13)

International Summit on Human gene Editing: A Global Discussions. 2015. http://nationalacademies.org/gene-editing/Gene-Edit-Summit/

Norrgard, Karen. 2008. "Genetic counseling: genetic testing, family history and psychosocial evaluation." *Nature Education* 1(1): 36

UNESCO. *25 Recommendations on the ethical, legal and social implications of genetic testing. 2004.* http://www.unesco.org.tr/dokumanlar/biyoetik_komitesi/25%20recommendations%20ec.pdf

Savulescu, Julian, Jonathan Pugh, Thomas Douglas, and Christopher Gyngell. 2015. "The moral imperative to continue gene editing research on human embryos." *Protein & cell* 6.7: 476-479.

Stolberg, Sheryl Gay. 1999. "A Death Puts Gene Therapy Under Increasing Scrutiny." *New York Times* (November 4).

Melnov Serghey B. 2016. "Bioethical problems of human genetics". Глобальная биоэтика в социальном измерении: материалы научно-практической конјеренции, Минск, 16—18 декабря 2015, Международный государственный экологический институт им. А. Д. Сахарова БГУ: к 70-летию ЮНЕСКО: 217-227.

Emilian Mihailov

6 Refocusing the Nudge Debate on Organ Donation

6.1 Introduction

In ideal circumstances, five patients can be saved by organ transplantation from just one deceased donor. Thus, a much smaller number of donors compared with the number of patients would solve the medical need. However, in the real world, there is severe organ shortage. On average, 16 patients die every day waiting for the organs they need (*Health-EU Newsletter* 183). There are many proposals to improve organ availability, from changing practices regarding end-of-life care to supporting cutting-edge research into interspecies chimeras. Improvements could be seen if patients are given the choice to donate their organs in the event of withdrawal of life support in intensive care or assisted suicide (Wilkinson and Savulescu 2012; Shaw 2014). Recent developments in gene editing might allow human organ generation in animals whose organ size, anatomy, and physiology are closer to humans (Wu et al. 2017).

But the most entertained policy change is the shift from an opt-in system to an opt-out system of organ donation, because it promises a big impact at the cost of small changes (Thaler and Sunstein 2009; Halpern 2016). Switching the *status quo* from registering into a potential donor list to registering out of a potential donor list, is expected to bring significant increases in donation rates. In opt-in systems, people have to act themselves, thus spending time and other resources, to register. In opt-out systems, people are presumed potential donors by default. If they wish to donate, they just do nothing. If they do not wish so, then they could actively express their unwillingness. This contrast is framed as informed, explicit, or positive consent versus presumed or implicit consent.

The shift toward presumed donorship is inspired by research on nudges from behavioral sciences. A nudge is a particular design on how people's possible choices are arranged. We nudge someone when we set up his/her choice context so as to increase the likelihood of picking choice (A) over choice (B), even though it would still be easy to choose (B). For example, if we want to increase the consumption of a food item, we display it at the eye level, while other products will be displayed last. Private companies nudge clients all the time, but governments have started to do this as well. The most famous examples are the nudge units inside the US and UK governments, which draw on social and behavioral sciences to influence people's decisions in directions considered by themselves better off or conducive to preserving public goods (Sunstein 2014; Halpern 2016). These teams of social scientists identify low-cost policy tools to nudge people to make more savings, eat healthier, or reduce pollution. Proponents of nudging as means of increasing organ donation point to huge differences in consent rates between policies that presume people to be donors and policies that do not. The promise of nudging to meet the medical need of patients with organ failure can explain why so many countries are increasingly attracted to opt-out legislation.

Nevertheless, the insistence on influencing people into preferring the *status quo* of a potential donor has been met with critical reactions of curtailing people's autonomy (MacKay and Robinson 2016; Rodrıguez-Arias and Morgan 2016). Although the opt-out policy appeals to consent language, it may undermine the standard of autonomy asserted by explicit consent. As Janet Radcliffe Richards puts it, "If you think we should accept as a *fundamental right* that organs should not be used without positive consent, that in itself settles the question. (...) On the other hand, if you do not think there should be such a principle, the question about policy remains open" (2013, 155).

In this paper, I argue for refocusing the nudging potential to increase organ donation as it is not inherently opposed to consent standards. The debate is too much focused on a type of intervention that raises worries about respecting people's autonomy and does not have a great-sized effect to justify the insistence on opt-out schemes. As I will show, nudges are versatile enough in their *modus operandi* to mitigate concerns about respect for autonomy and have a much wider application. Firstly, I present what motivates policymakers to push for opt-out legislation. Then I will make the case for a more realistic picture about the impact of such systems, downplaying their attraction. Thirdly, I will argue that the current opt-out legislation is based on dubious usage of consent standards and that the nudging potential can be refocused within the framework of explicit consent. In the end, I will further suggest how behavioral interventions can be rerouted to have a wider application. We can also nudge pivotal stakeholders such as family members and medical professionals who act on behalf of public institutions, not only potential donors.

6.2 The success of opt-out systems

In general, people are willing to help. A recent poll shows that 95% of Americans strongly support organ donation (National Survey of Organ Donation Attitudes and Behaviors 2013). According to a wide public opinion report in 2009, 55% of Europeans are willing to donate one of their organs to an organ donation service immediately after death (Special Eurobarometer 333a 2010). If we break down these results by individual countries, some show even greater support. In countries such as Sweden, Malta, Belgium, Finland, and Denmark, >70% are willing to donate. The gap between support and action poses serious challenges to policymaking.

What kind of measures are effective in moving people from approval to action? In answering this difficult empirical question, the fact that there are huge differences in the number of registrations between countries is often pointed out. Why are only 28%

of Americans but 99.9% of French citizens registered as donors? Why are only 4.25% of Danish citizens but 85.6% of Swedes registered as donors, especially since both strongly support organ donation? It is not very convincing to think, as Gigerenzer puts it, that Americans are "more anxious about a post-mortem opening of their bodies than the French" (2008, 2). As previously highlighted, attitude surveys do show a generalized willingness to donate.

These numbers are better explained by differences in default positions. In Hungary, France, and Sweden, the policy default is that everyone is presumed to be a donor, whereas in Denmark, the US, and England, the default is that nobody is presumed to be a donor. Thus, in opt-out systems, people have to explicitly express their preference not to donate and in opt-in systems, they have to explicitly express their preference to donate. What is at work here is the behavioral rule: "If there is a default, do nothing about it" (Gigerenzer 2008). This rule can be overridden if preferences count against it and the costs are low. Only 0.1% of French citizens and 0.075% of Polish citizens opted out from donor default, which might indicate that for many, there is almost no preference against having donor status. Only 17% of British citizens opted out from nondonor default. This could indicate that only those with strong preferences of becoming donors did something about it. If people do not bother that much to alter a default, then probably not too many will sign a donor card when they are nondonors by default and not too many will do something to become nondonors. It seems that defaults can influence choices in at least two ways (Johnson and Goldstein 2003). On the one hand, defaults can be perceived as reasonable recommendations to save human lives, and, on the other hand, accepting a default is effortless. Depending on the specifics of registration, an active decision can be unpleasant and time consuming.[1]

In their famous study, Johnson and Goldstein (2003) confirmed the powerful effect of defaults on agreement rates. They asked participants whether they would be donors, varying defaults in framing the questions. In the opt-in condition, participants entertained the scenario that they moved to a new country where the default was to be nondonor, and they were given a choice to confirm or change that status. In the opt-out condition, the wording was identical but the choice was to confirm or change the status of donor. In the neutral condition, subjects had to choose with no prior default. The results showed the highest donation rates for the opting-out condition, almost twice as high (82%) in comparison with the opting-in condition (42%). Even the neutral condition was significantly higher (79%) than the nondonor default.[2]

1 In Romania, for instance, the process is burdensome. Firstly, one has to notarize a donation agreement, and then file identification data, as well as data about the notary who attested the donation agreement, to the National Register of Donors (Ministry of Health Order 1158/2012).
2 The results were replicated by Vladucu et al. (2016), showing the highest rates for the opting-out condition, followed by the neutral condition.

Recently, an opt-out law came into force in Wales, providing a positive case for default effects. Of the 60 organs transplanted between December 2015 and June 2016, half came from people who had not objected to opt out. In the same period in 2014–2015, before the law came into force, 23 people donated their organs, while in 2013–2014, only 21 donated.[3] More generally, Abadie and Gay (2006) analyzed the impact of opt-out legislation in 22 countries over a 10-year period. After controlling for other factors that contribute to organ donation, they found that switching from opt-in to opt-out systems increases donation rate by approximately 16%.[4] Similarly, Shepherd et al. (2014), analyzing a much bigger sample of 48 countries (23 opt-in and 25 opt-out models) over a 12-year period, found that overall opt-out consent is associated with greater deceased donor rates. Considering this potential to increase donation, many medical associations and government representatives have called for opt-out legislation.

6.3 Tempering the opt-out enthusiasm

It is understandably tempting to highlight huge differences in donation agreements, which are mainly due to fairly simple changes in default status. What should we prefer: a policy with an impact of 20% registered donors, or one with over 85%? When the choice is framed in this way, the answer is obvious.

Indeed, default policies increase the number of registered donors, but to have an impact on organ shortage more is required, because agreement rates do not necessarily translate into donation rates. Leaving aside the issue of quality, as the aim is also to make available organs of above average quality[5], what matters most is the increase in organ donation after death or transplanted patients. When someone makes an active choice to register, this could be taken as expressing a strong and unambiguous willingness to donate. However, there is no guarantee that the donor's organs are suitable or that the family will not veto donation wishes. The problem of translating agreement rates into donation rates is more stringent when people are presumed donors by default. Passive agreements are ambiguous with reference to willingness to donate. When someone does not object to presumed donor status, this could be due to a willingness to donate, but it could also mean a lack of status awareness in the first place or a lack of interest for remote events. Additionally,

3 For more details, look up https://www.organdonation.nhs.uk/statistics/.

4 Using a multiple regression analysis, Johnson and Goldstein (2003) obtained roughly smaller results.

5 Available organs are many times below average quality. This has cumbersome implications for resource allocation policy, whether to use higher-risk organs in the more-urgent-need patients or to use higher-risk organs in the less–urgent-need patients.

accepting a default is more often than not effortless, while changing it involves some costs. So, a decision not to object may be the result of personal comfort. The opacity of passive agreements with reference to donation preferences leaves room for family involvement to steer the decision either way. As a result, increases in consent rates through presumed agreements may not tell us much about what to expect with reference to actual donations. For example, some findings from experimental studies showed that nudge interventions increased registration rates, without affecting actual donation (Farrell 2015). Similarly, the presumed consent law that came into effect in Japan in 2010 has not increased numbers as expected. The rate of deceased organ donations has remained roughly the same even after the revision of the law (Soyama and Eguchi 2016).

Differences in registered donors between opt-out and opt-in legislations can be misleading and perceived as artificial if there is no significant difference in donated organs as well. Once we distinguish between donor status and donation *per se*, we get a messy picture about the impact of default policies on donation rates (Figure 1). There is no pattern suggesting that countries with presumed consent automatically outperform countries with explicit consent.

Legislation type	Country	2012 rate p.m.p.	2013 rate p.m.p.	2014 rate p.m.p.	2015 rate p.m.p.
Opt out	Spain	87.6	88.8	90.2	100.7
Opt out	Hungary	34.1	38	52.4	48.4
Opt out	Belgium	93.9	87.9	83.4	87.3
Opt out	Italy	51.1	50.3	53.2	55.6
Opt out	Finland	55.7	50.7	62.4	67.8
Opt out	France	76.3	76.5	79.6	85.8
Opt in	The Netherlands	73.8	72.9	78.3	74.7
Opt in	Denmark	56.6	53.8	63.6	68.8
Opt in	England	65.1	73.2	71.8	69.6
Opt in	Germany	53.3	47.7	44.9	45.5
Opt in	Romania	11.9	19.2	20.3	17.7
Opt in	Bulgaria	2.6	5.4	11	10.0

Figure 1. Annual rate of total number of patients transplanted. Data extracted from *European Council Newsletter Transplant* Vol. 21, 20, 19, 18).

Spain and Belgium, countries with opt-out legislation, have almost twice the donation rate of Germany and Denmark (opt-in legislation), but the Netherlands and England (opt-in models) have higher donation rates than Hungary, Italy, and Finland (opt-out models). Significant differences can also be seen between countries with the same legislation. For example, the Netherlands clearly outperforms other opt-in countries, while Spain has the highest donation rate among all opt-out countries. It seems that there are other background factors that explain these variations.

When we consider donation *per se*, it does not help too much to compare the registration rates of countries that have implemented an opt-out system with ones that have not. The variability of transplanted patients must be explained by other background factors that contribute to the overall outcome, such as medical and transplant infrastructure, special training, education level, propensity to donate or social norms, and religious beliefs about the use of postmortem bodies. Although it has an opt-in system, the US does better than other opt-out countries because of its superior medical system in matching donors with recipients, delivering the organs, and performing successful transplants (Thaler and Sunstein 2009).[6] Spain has become the world leader mainly due to organizational measures. The Spanish model consists of earlier referral of possible donors to transplant coordination teams, a benchmarking project to identify critical success factors in donation after brain death, new family approach and care methods, the development of training courses aimed at specific groups of professionals, as well as national strategic plans (Matesanz et al. 2011).

Besides the relevance of medical infrastructure and organizational settings, there is a truth of the matter hard to digest and often omitted in public debate. Countries with low donation rates have, in general, a low mortality rate relevant for organ donation, while countries with high donation rates have higher mortality rates (Coppen et al. 2005).[7] When controlled for mortality rates, there is no automatic superiority of opting-out systems. As it happens in organ donation, one man's tragedy is another man's salvation.

Moreover, there is a variability in policy application. The majority of countries with opt-out legislation, including the highest performer Spain, has adopted in practice what has been called a soft version of presumed consent. Doctors are still asking for family consent, even if the potential donor had not objected to presumed status during his/her lifetime (Rithalia et al. 2009). By contrast, in the hard version of presumed consent, organs can be used without family consultation. The lack of objection from a potential donor during his/her lifetime to the presumed donor status is sufficient. Now, if doctors still ask for family consent in presumed consent systems, this raises the question to what extent a soft version of opt-out systems understates the impact of defaults, as its objective is to bypass active choice as a means to increase organ

6 The reduced medical and transplant infrastructure could be a critical factor that explains why developing countries such as Romania and Bulgaria have low donation rates.

7 This raises a practical dilemma with thorny trade-offs. What should we do if the number of available organs are dependent on high mortality rates? Should we adopt policies that reduce the number of car accidents, for instance, but then we will have low donation rates?

donation? Currently, there are not enough data to compare soft and hard versions of opt-out systems (Shepherd et al. 2014). To elucidate what factors contribute decisively to donated organs, future research will have to shed light on this issue.[8]

We should be cautious with predictions that opt-out legislation would increase organ donation by roughly 15%. Analyses need to consider more relevant factors and how they interact in practice. The potential of opt-out systems to increase organ donation needs to be qualified in many respects if we want to avoid exaggerated expectations. It is overoptimistic to insist that small changes in default positions can have a big impact.

6.4 Opt-out policy and consent standards

What gives weight to opt-out policy is the argument that switching the default position from nondonor to donor can make a big difference in reducing organ shortage. Even if the opt-out policy may fall short for other ethical standards, it could still be regarded as morally permissible on grounds of promoting beneficence. In health-care settings, dilemmas in which one has to balance competing principles of beneficence and respect for autonomy are pervasive. This poses the question of how incompatible are defaults with consent standards.

Although explicit and presumed consent policies are different, Thaler and Sunstein argue that opting-out preserves freedom of choice by giving the opportunity to easily object (2009, 187). However, from the fact that a policy preserves freedom of choice, an adherence to consent requirements does not follow. Consent is thought to involve more than preserving freedom of choice. Imagine a company that subscribes you each day to a news magazine and, at the same time, it allows you to easily object to subscriber status. In this scenario, we still have the freedom to change the status. However, it does not follow that consent standards are being met, as they also require, in different contexts, seeking prior understanding from a subject or authorization for an intervention. The subscription case involves the freedom of opting out, but not an initial authorization of enlisting.

Describing opting-out policies as being based on presumed, implicit, or tacit consent is also confusing. These forms of consent make sense in particular contexts that we need to be aware of if we do not want to carry consent language too far. In clinical practice, consent is usually presumed on at least two conditions: a particular

8 At first sight, there seems to be very little difference in impact between opting-out and opting-in schemes as long as family members have in practice the power to veto. I suspect there is an important difference even if doctors are still asking for family consent in opt-out systems. This is because nudges are primarily about designing a choice context. The choice of seeking family consent is inescapable in opt-out systems, whereas in opt-in systems, doctors could easily side step the donation option.

person *cannot* explicitly consent (for different reasons) and we presume something on the basis of *available* information about the personal history of that particular person.[9] Implicit consent is inferred from *previous* commitments. When patients consent to surgery, it is also implied that they consent to other procedures that have not been specifically named in advance but which are necessary for the success of the medical intervention. Tacit consent is expressed when someone *knowingly* accepts a state of affairs by omitting to do something about it.

Given these standard forms of consent, the opt-out policy seems to be related to tacit consent, rather than to presumed or implicit consent. The attribution of donor default is not done on the basis of available information about the values and preferences of citizens, nor is it inferred from their previous commitments. It is more like assigning a status that is tacitly confirmed by omitting to object.[10] But even this framing of consent is problematic since tacit consent is dependent on prior knowledge of the state of affairs, which is generated by omitting to act. The opt-out legislation does not satisfy this condition. If the lack of objecting to a donor default is to express tacit consent, then we must have evidence that all citizens are aware and understand the policy. However, obtaining such evidence is extremely hard. Public information campaigns can raise awareness, but they cannot guarantee, on a countrywide scale, the standards that are needed for valid tacit consent.[11]

Current opt-out legislations are based on loose talk of consent as it does not align well with standard requirements of presumed, implicit, and tacit consent. And the fact that opting out opposes explicit consent is considered sufficient to settle its moral permissibility. The complicated empirical question about its impact on donation rates will have to be bracketed if the opt-out policy fails the normative test of explicit consent. There is a strong intuition that at least in health-care settings "consent should refer to an individual's actual choices, not to presumptions about the choices the individual would or should make" (Beauchamp and Childress 2001, 66). The use of the opt-out policy is widely viewed as morally suspicious, especially because it bypasses our rational capacities and actual choices (Saghai 2013; MacKay and Robinson 2016; Levy 2017).

---

9 Consent is also thought to be presumed on the basis of a theory of rationality, labeled in philosophical ethics as rational consent. This way of thinking about presumed consent is not influential in clinical practice.

10 Some defenders of the opt-out scheme admit that the label "presumed consent" is confusing but insist on justifying it by appealing to implicit consent (refer Saunders 2012).

11 It remains an open question whether on a countrywide scale, the standards of evidence for tacit consent should be different from those for individual cases. One could argue that population surveys that show awareness over an absolute majority threshold constitute evidence for consent. The issue is whether it properly counts as sufficient evidence for consent or more is needed.

6.5 Beyond the dichotomy of presumed and explicit consent

The insistence on nudges that capitalize on attributing a presumed status without prior interaction encourages a misleading picture. Once it is clarified that nudges are primarily concerned with the architecture of choice, we see that behavioral interventions are not inherently opposed to explicit consent. The focus on opt-out forms of intervention overlooks the diversity and potential versatility of nudges to comply with consent requirements, encourage autonomous decisions, engage deliberative capacities, and reduce the costs of active choice.

We often do not form preferences on particular issues simply because they are not close enough to our immediate choice context. The question of what could happen with our organs after we die or whether we are saving enough for retirement is not something that grips our daily lives. Most probably we do not bother that much if there is no urgency. Add to this our daily inertia and routines, and it is not surprising that people fail to act, even according to their own preferences. This is where choice architects intervene. The aim is to make choosing easier when it is time consuming or cognitively demanding in searching for information. The strategy is to intervene in how people's possible choices are arranged. We nudge someone when we arrange his/her choice context in order to influence the likelihood of choosing option (A) over option (B), even though it would still be easy to choose (B), and (A) is not worse than (B). We also nudge someone when we simply introduce options that are rarely considered in his/her choice context.

This is not incompatible with explicit consent requirements because nudging interventions do not necessarily have to bypass actual choices in order to influence decision-making. The only dissimilarity between conditions in the study by Johnson and Goldstein was not presumed consent versus explicit consent, but the contrast between confirming a donor status and confirming a nondonor status. Thus, reasons that explain nudging effects could equally work within an explicit consent framework. A nudge is supposed to work because decision-makers perceive defaults as reasonable recommendations. If the perception of reasonable recommendations is what sways people to accept the *status quo*, then it does not matter that much how we frame the consent requirements. Instead of waiting for people to object to donor status, we could explicitly ask them, when they interact with public institutions, whether they would like to confirm or change the *status quo*.[12]In addition, filling out many forms or perceived bureaucracy can discourage people to sign in as donors, while accepting a default is effortless. Consequently, we achieve a nudging effect when we make registration as easy as possible, ideally at a mouse click.

12 Default effects persist even when people are informed of the presence of default options (Loewenstein et al. 2015).

Note that in Johnson and Goldstein's study, the neutral condition, which does not stipulate a default option, was just as effective in obtaining consent rates (79%) as the opt-out condition (82%), compared with the opt-in condition (42%). "Neutral condition" is a misleading label. It makes us overlook its nudging effect, which works by intervening in a choice context just to introduce options that are not of immediate concern for people in their daily life. This implies that the consent rate can be improved by presenting people a donation option when, for instance, they get a driver license, change their ID, or undergo a routine medical checkup.

Another tool of choice architecture compatible with explicit consent is the provision of relevant information that does not usually lie around. Thaler and Sunstein (2009, 190) describe the case of Illinois in the US. In promoting organ donation, the State of Illinois highlighted how many people are on the waiting list, as well as the fact that 87% of adults in Illinois believe that registering as an organ donor is the right thing to do. In the same manner, the introduction of a donation option in choice contexts can go together with information about the importance of the problem, evaluative attitudes, and results. This tool encourages autonomous decisions by providing relevant information. It assumes that, when given the possibility, people will lean more toward informed choices, understood as responsive to relevant facts, and morally desirable outcomes.

Furthermore, there are cases in which questionable beliefs negatively influence the willingness to donate. In Turkey, people are most likely to refuse organ donation for religious reasons (29%). Other countries where people are more likely to be reluctant to donate organs for religious reasons are Romania (17%), Austria (15%), Macedonia (12%), and Slovakia (11%).[13] But what is striking is that the religious beliefs are contrary to positions held by the Orthodox and Catholic churches, as well as by Islam. All support organ donation with therapeutic aims, as a sign of kindness and love for fellow human beings. In such cases, a donation option could be presented along with information about the support from religious institutions, leading figures, or holy texts. Nudges such as these can make people self-conscious regarding the epistemic status of their religious beliefs. The hope of providing relevant information and prompting deliberative capacities is a revision of spurious beliefs. This type of intervention can be applied to family interviews, as well as other behavioral interventions that nudge families toward more deliberative decisions.[14]

13 Among Europeans, fear of manipulation of the human body and distrust in the system are the most prevailing reasons for not donating organs (refer Special Eurobarometer 333a 2010, 26).

14 A study in a French organ procurement centre (Le Nobin et al. 2014) documented that the most frequent reason for family refusal was the desire to keep the body's wholeness (46.3%), followed by religious opposition (16.4%). Doctors could make family members aware of false beliefs, cognitive bias, and misunderstanding and thereby shift involved parties toward a more-reflective frame of mind (Shaw and Elger 2014; Stefanescu Schmidt et al. 2017).

6.6 Nudging beyond potential donors

Dubious talk of consent relating to the opting-out policy and its contrast with explicit agreement fueled charges that it poses threats to people's autonomy. But as I have argued, nudging interventions can be made compatible with explicit consent requirements. Rerouting nudges away from presumed consent could also have benefits for family involvement in postmortem decision-making. Because presumed consent is ambiguous with reference to willingness to donate, obtaining actual choices to donate could provide more moral authority to families to fulfill one's wishes.

Refocusing the potential of behavioral insights can go even further than private citizens and family members. How medical institutions perform is essential for increasing donation rates. Ultimately, nudging advocates believe in better governance (Thaler and Sunstein 2009; Halpern 2016). The moral advantage of targeting individuals acting on behalf of public institutions is that concerns applicable to private citizens do not hold for public servants. Doctors, by their very professional identity and institutional context, are already committed to forgo options, which are at liberty for private individuals, and accept interventions to achieve public goods.

So, it is natural to look more into how we could nudge relevant parties to perform better according of their own mission. Applications of behavioral insights from other medical challenges can serve as paradigms that could guide future research. For instance, there is a widespread use of antibiotics to treat infections for which it has little clinical advantage, killing beneficial bacteria as well (Steinman et al. 2003; Blaser 2011). Many interventions have been used to reduce unnecessary prescription of antibiotics, such as physician and patient education or computerized clinical decision support, but in a recent investigation, Meeker et al. (2016) implemented insights from behavioral research in primary care practices. One nudging intervention was peer comparison. Clinicians with the lowest unnecessary prescribing rates were informed monthly that they were "top performers". Clinicians with higher rates received monthly e-mails, which included the number and proportion of their antibiotic prescriptions for antibiotic-inappropriate infections, alongside the proportion of top performers. Compared with baseline period, the intervention period reduced the rate of unnecessary prescriptions from 35% to 19.2%. In another randomized controlled trial, social norm feedback from a high-profile messenger substantially reduced antibiotic prescribing (Hallsworth et al. 2016). The intervention consisted of sending a letter from England's Chief Medical Officer and a leaflet on antibiotics for use with patients to general practitioners. The letter described how the practice was prescribing antibiotics at a higher rate than 80% of practices in its local area.

Similarly, donation rates could be increased by nudging specific groups of professionals through interventions such as peer comparison and social norm feedback from a high-profile messenger. Indeed, the issue of overuse of antibiotics and that of organ shortage are quite different. Antibiotic prescription is incumbent on the doctor's decision, whereas organ donation is not. While it is true that because of this asymmetry, we cannot expect similar results, the central point of these examples is to illustrate how nudging applications could potentially target individuals acting on behalf of public institutions as well. It is a mistake to think that the contribution of medical professionals to increase organ donation is not subject to unwanted institutional inertia, social influences, or cognitive biases.

6.7 Conclusion

In effect, 99.9% of French citizens are presumed donors, compared to 4.25% of Danish citizens who have actively chosen to become donors. Understandably, the proposal to move toward opting-out systems is attractive, as it promises a big impact with small changes in default positions. However, these huge differences in registration rates are potentially misleading. Countries with opt-out schemes do not automatically outperform those with opt-in schemes in terms of actual donation rates. The contribution of opt-out systems to increase organ donation needs to be qualified in many aspects if we want to avoid false expectations. Additionally, there are normative issues. Current opt-out legislations are based on confusing talk of presumed, implicit, or tacit consent. The attribution of donor default is neither done on the basis of available information about the values and preferences of citizens, inferred from their previous commitments, nor is it dependent on prior knowledge of the state of affairs that is generated by omitting to act. And, especially, the fact that opting-out bypasses explicit consent is widely considered a decisive reason against it. But we need not use nudges only as tools of bypassing actual choices. The insistence on opt-out forms of intervention overlooks the diversity and versatility of nudges to comply with explicit consent standards and encourage autonomous decisions. Behavioral interventions can be refocused to make the actual choosing of potential donors easier when it is time consuming and cognitively demanding, and when the options are outside people's normal choice architecture. These interventions can have a much wider application, such as nudging family members toward more deliberative decisions during consent interviews, as well as inducing medical professionals to perform better as per their own mission.[15]

References

Abadie, Alberto, and Sebastien Gay. 2006. "The impact of presumed consent legislation on cadaveric organ donation: a cross-country study." *Journal of health economics* 25(4): 599-620.

15 I thank Mihaela Frunză, Dorina Pătrunsu, Reinhard Merkel, Blanca Rodríguez López, Emanuel Socaciu, Constantin Vică, Cristina Voinea, Toni Gibea, and Wangmo Tenzin for comments that improved this chapter.

Beauchamp, Tom, and James Childress. 2001. *Principles of biomedical ethics*. New York: Oxford University Press.

Blaser, Martin. 2011. "Antibiotic overuse: stop the killing of beneficial bacteria." *Nature* 476(7361): 393-394.

Coppen, Remco, Friele, Roland D., Marquet, Richard L., and Sjef K. M. Gevers. 2005. "Opting-out systems: No guarantee for higher donation rates." *Transplant international*, 18(11), 1275-1279.

Farrell, Anne-Maree. 2015. "Addressing organ shortage: are nudges the way forward?" *Law, Innovation and Technology* 7(2): 253-282.

Gigerenzer, Gerd. 2008. "Moral intuition = fast and frugal heuristics?" In W. Sinnott-Armstrong (Ed.), *Moral Psychology: Vol 2. The cognitive science of morality: Intuition and diversity*. Cambridge, Massachusetts: MIT Press.

Halpern, David. 2016. *Inside the Nudge Unit: How small changes can make a big difference*. London: Penguin Random House

Hallsworth, Michael, Chadborn, Tim, Sallis, Anna, Sanders, Michael, Berry, Daniel, Greaves, Felix, Clements, Lara, and Sally C. Davies. 2016. "Provision of social norm feedback to high prescribers of antibiotics in general practice: a pragmatic national randomised controlled trial." *The Lancet*, 387(10029): 1743-1752.

Johnson, Eric, and Daniel Goldstein. 2003. "Do defaults save lives?" *Science* 302(5649): 1338-1339.

Le Nobin, Julien, Pruvot, Francois-Rene, Villers, Arnauld, Flamand, Vincent, and Sebastien Bouye. 2014. "Family refusal of organ donation: a retrospective study in a French organ procurement center." *Progres en urologie: journal de l'Association francaise d'urologie et de la Societe francaise d'urologie* 24(5): 282-287.

Levy, Neil. 2017. "Nudges in a post-truth world". *Journal of medical ethics* 43:495-500.

Loewenstein, George, Bryce, Cindy, Hagmann, David, and Sachin Rajpal. 2015. "Warning: You are about to be nudged." *Behavioral Science & Policy* 1(1): 35–42.

Matesanz, Rafael, Domínguez-Gil, Beatriz, Coll, Elisabeth, de la Rosa, Gloria, and Rosario Marazuela. 2011. "Spanish experience as a leading country: what kind of measures were taken?" *Transplant International* 24(4): 333-343.

MacKay, Douglas, and Alexandra Robinson. 2016. "The ethics of organ donor registration policies: Nudges and respect for autonomy." *The American Journal of Bioethics* 16(11): 3-12.

Meeker, Daniella, Linder JA, Fox CR, Friedberg MW, Persell SD, Goldstein NJ, Knight TK, Hay JW, and Doctor JN. 2016. "Effect of behavioral interventions on inappropriate antibiotic prescribing among primary care practices: a randomized clinical trial." *JAMA* 315(6): 562-570.

Newsletter Transplant Vol. 21, No 1 Sept. 2016.

Newsletter Transplant Vol. 20, No 1 Sept. 2015.

Newsletter Transplant Vol. 19, No 1 Sept. 2014.

Newsletter Transplant Vol. 18, No 1 Sept. 2013.

Richards, Janet Radcliffe. 2012. *The ethics of transplants: Why careless thought costs lives*. New York: Oxford University Press.

Rithalia, Amber, McDaid, Catriona, Suekarran, Sara, Myers, Lindsey, and Amanda Sowden. 2009. "Impact of presumed consent for organ donation on donation rates: a systematic review." *BMJ* 338: a3162.

Rodrıguez-Arias, David, and Myfanwy Morgan. 2016. "'Nudging' Deceased Donation Through an Opt-Out System: A Libertarian Approach or Manipulation?" *The American Journal of Bioethics* 16(11): 25-28.

Romanian Ministry of Health Order 1158/2012.

Saghai, Yashar. 2013. "Salvaging the concept of nudge." *Journal of medical ethics* 39(8): 487-493.

Saunders, Ben. 2012. "Opt-out organ donation without presumptions." *Journal of medical ethics* 38(2): 69-72.

Shaw, David. 2014. “Organ donation after assisted suicide: a potential solution to the organ scarcity problem.” *Transplantation* 98(3): 247-251.

Shaw, David, and Bernice Elger. 2014. “Persuading bereaved families to permit organ donation.” *Intensive care medicine* 40(1): 96-98.

Shepherd, Lee, O’Carroll, Ronan E., and Eamonn Ferguson. 2014. “An international comparison of deceased and living organ donation/transplant rates in opt-in and opt-out systems: a panel study.” *BMC medicine* 12(1): 131.

Soyama, Akihiko, and Susumu Eguchi. 2016. “The current status and future perspectives of organ donation in Japan: learning from the systems in other countries.” *Surgery today* 46(4): 387-392.

Special Eurobarometer 333a. 2010. *Organ donation and transplantation*.

Stefanescu Schmidt, Ada C. Bhatt, Ami and Cass Sunstein R. 2017. “Boundedly Rational Patients? Health and Patient Mistakes in a Behavioral Framework.” Available at SSRN: https://ssrn.com/abstract=2953250

Steinman, Michael A., Gonzales, Ralph, Linder, Jeffrey A., and C. Seth Landefeld. 2003. “Changing use of antibiotics in community-based outpatient practice, 1991–1999.” *Annals of Internal Medicine* 138(7): 525-533.

Sunstein, Cass. 2014. “Nudging: a very short guide.” *Journal of Consumer Policy* 37(4): 583-588.

Thaler, Richard, and Cass Sunstein. 2009. *Nudge: Improving decisions about health, wealth, and happiness*. London: Penguin Books.

Vlăducu, Andrei, Diaconu, D, Dobre, Dani, and B. Deaconu. 2016. “Status-quo și donarea de organe în România. Un studiu descriptiv cantitativ.” *Revista de Psihologie* 62(1): 45-56.

U.S. Department of Health and Human Services. 2013. *National Survey of Organ Donation Attitudes and Behaviors*.

Wilkinson, Dominic, and Julian Savulescu. 2012. “Should we allow organ donation euthanasia? Alternatives for maximizing the number and quality of organs for transplantation.” *Bioethics* 26(1): 32-48.

Wu, Jun, Platero-Luengo, Aida, Sakurai, Masahiro, Sugawara, Atsushi, Gil, Maria Antonia, Yamauchi, Takayoshi, Suzuki, Keiichiro, Bogliotti, Yanina Soledad, Cuello, Cristina, and Mariana Morales Valencia. 2017. “Interspecies Chimerism with Mammalian Pluripotent Stem Cells.” *Cell* 168(3): 473-486.

Mihaela Frunză, Iulia Grad, Sandu Frunză, Ovidiu Grad

7 CLICK HERE! To Find More About Organ Transplantation: Ethical Aspects of Media Stories on Organ Donation from Romanian Newspapers

7.1 Introduction

Discussion concerning organ donation and transplantation is always a conversation of generosity because the topic encompasses the idea of a gift (Frunză et al. 2010). Gifts situate us in a place of the opening toward each other, offering us the opportunity for profound reflection about what we are, the meaning of our lives, and the resources available to save and improve the lives of others.

As such, careful discourse about donation and transplantation is a necessary part of overcoming the ethical dilemmas surrounding organ donation decisions. This debate is needed regardless of whether we speak about the donors or their families who are tasked with making decisions on their behalf. The discussion also involves those who perform the medical procedures, examining everything from their professional responsibilities to the ethics of their social responsibilities (Frunză 2011).

Organ transplantation is a life-saving procedure whose full potential, unlike other medical procedures, depends on the public's willingness (which encompasses the willingness of donors and of their next-of-kin) to provide organs for patients in need. Therefore, the "social acceptability" of medical strategies that result in public acceptance is crucial in the establishment of an effective national transplant system, as are the proper medical infrastructure and qualified medical professionals (Rodriguez-Arias 2013). Traditional and nontraditional media play a central role in mobilizing the public because organ donation is not a frequent topic of conversation in Romania[1].

Although organs may be donated by either living or deceased donors, the focus of this article will be on deceased donation; due to the transplant risks for living donors, the communication strategies that might be used to mobilize them are very different. However, while we do not want to minimize the contributions of living donors, it is a separate topic that must be addressed with a different set of research tools.

---

1 Some studies suggest that even if organ donation would be present more often in daily conversations, affirming one's opinions on donation does not guarantee that they will be implemented by one's family, in the event of a sudden death (Holman et al. 2013).

Mass media campaigns can promote positive changes or prevent negative changes in health-related behaviors (Wakefield et al. 2010). Public communication campaigns that aim to encourage organ donations by deceased persons are carried out all over Europe. These campaigns aim to inform the general public about the benefits of organ donation and how future donors can register, depending on the type of consent[2] required in the national context. Even if the long-term impact of organ donation campaigns is difficult to accurately estimate, their relevance in terms of promulgated information has been well-documented across Europe as well as globally (Krekula et al. 2009; Feely and Moon 2009; Morgan and Harrison 2010).

Until these campaigns are effectively organized in Romania, the general public will use traditional media as the primary source of information on organ transplantation (Karner-Huțuleac 2012; Ioan et al. 2011). Therefore, media depiction of transplantation is one of the factors that shapes peoples' attitudes and behaviors (Ioan et al. 2011; Holman et al. 2013). For instance, the media may perpetuate the existing urban myths and stereotypes on transplantation, or it may act to better inform the public about organ donation in general and the importance of increasing the overall donation rates.

Traditional media, particularly paper-based media (newspapers and periodical magazines), has been shaken by significant crises in recent years. These scandals affect the perceived legitimacy, circulation, number of copies sold, and public impact. For the surviving periodicals, these events pushed them toward tabloidization, commercialization, and the ever-increasing emphasis on "infotainment". Nevertheless, paper-based media remains an important outlet for the dissemination of mass-scale useful information, such as knowledge concerning the national transplant system and organ donation.

The aim of our paper is to offer the readers insight into the way organ donation and transplantation have been depicted in the Romanian media. Thus, our paper draws on previous research that the members of our team have performed on the subject matter: namely, the analysis of the depictions of transplantation within Romanian media over the past decade, which is presented in the section "Sensational stories on organ donation in the Romanian paper-based media (2008–2012)" (Frunză et al. 2011; Holman et al. 2012). In addition, we have performed a new analysis for the year 2015 (January–December), which is detailed in the section "'Look, mom, now I've got hands!'. Unlikely Stories of Transplantation from the Recent Romanian Periodicals".

2 Generally speaking, there are three types of default solutions that are used worldwide when registering consent to organ donation: explicit consent (when the default position is the nondonation and the would-be donors must actively register their donation intention while alive), presumed consent (when the default position is the donation and the people who do not wish to donate may register their refusal while alive), and mandated choice (when everybody is actively asked whether they wish to donate or not and their decision is registered) (Van Dalen and Henkens 2014).

In order to give the reader some positive, if limited, examples of what can be done at a small scale to correct the pessimistic image of transplants in the media, the section "From Mutual Ignorance to Mutual Partnerships: Students' Campaigns Promoting Organ Donation" presents examples of student-inspired campaigns to promote organ donation in Romania.

7.2 Sensational stories on organ donation in the Romanian paper-based media (2008–2012)

The year 2008 serves as a landmark in our analysis because in this year, there was an attempt to change the Romanian legislation on transplantation and to introduce presumed consent. This sharply diverges from the existing legislation that requires explicit informed consent by a member of the donor's family (Frunză et al. 2011). It is for this reason that we decided to monitor the articles published during 2008 (January–December) in two influential Romanian newspapers – *Adevărul* (www.adevarul.ro) and *Cotidianul* (www.cotidianul.ro). *Cotidianul* was selected because, apart from its wide circulation, it is the only periodical openly in favor of presumed consent (despite the public opinion leaning toward conserving the status quo).

Experts worldwide compared the relative efficacy of the two types of consent systems – explicit and presumed – and their results are subject to ongoing debates. Legislation promoting presumed consent has been singled out as a significant factor that increases the donation rate (Mossialos et al. 2008; Abadie and Gay 2006; Gimbel et al. 2003). In addition, presumed consent has been proved to improve the attitude toward donation in the case of undecided respondents (respondents who do not have a definite opinion concerning organ donation) (Van Dalen and Henkens 2014). Nevertheless, one of the most thorough meta-analysis on this comparison (Rithalia 2009) concluded that, despite the reportedly increased donation rates following the introduction of presumed consent, the type of consent is unlikely to be the only factor responsible for the high donation rates seen in certain countries.

However, even if presumed consent is not a magical recipe for increasing donation rates, the case study of Romania shows that it generated the first consistent public debate on transplant matters, with a notable argumentative and ethical component. In the premier study carried out in Romania, the authors studied the articles and the readers' comments from two national newspapers, *Adevărul* and *Cotidianul*, during the year 2008 (Frunză et al. 2011). In this respect, the two periodicals echoed the fate of the proposed bill in the Parliament (where the bill was finally rejected). From an ethical perspective, the authors note that there was a significant difference in argumentation between the two sides: those favoring the adoption of presumed consent provided either utilitarian or evidence-based views, whereas those favoring the status quo were highly emotional or hypothetical. Romania's ad hoc "rhetorical battle" was settled in favor of those opposing presumed consent.

Another study done in the Romanian context studied articles and user-generated comments from eight periodicals: three quality newspapers (*Adevărul*, *Gândul*, and *Evenimentul Zilei*), three tabloid newspapers, and two regional newspapers (*Monitorul de Cluj* and *Bună Ziua Iași*), followed during 2010 (January–December) (Holman et al. 2012; Holman et al. 2013). The analysis had an important quantitative dimension made possible through discursive analysis. There were plenty of ethical topics presented in the 309 articles published on transplant/organ donation in 2010. However, the apparent opening in articles with seemingly positive messages was rhetorically undermined by various factors, most significantly, the avoidance/underrepresentation of important issues such as consent or small donation rates.

7.3 "Look, mom, now I've got hands!" Unlikely stories of transplantation from recent Romanian periodicals

We have picked up for analysis articles and user-generated comments from one quality newspaper (*Adevărul*) and one tabloid newspaper (*Click!*) for the period spanning 2015, in order to provide comparable results with previous research (of 2008 and 2010). In the interest of consistency, we again selected *Adevărul* and, due to its high circulation, the tabloid *Click!* (also previously covered in the 2010 analysis). Our main focus remained on the public attitudes toward transplantation with special emphasis on the issue of consent for donation. We wanted to add a dimension that previous research had not accurately captured and that we deemed important in modern "videocratic" society, so we also explored the visual elements (both photos and videos included in the online edition of the newspapers) that accompanied the articles. A keyword search including relevant terms (organ transplantation, organ donation, and consent for donation) was performed in the archived data of the two newspapers (by author MF). Articles had to mention the medical procedure of transplantation in order to be selected[3]. Two researchers (authors MF and IG) read the content of each article and reached a consensus concerning their relevance for the research.

Table 1 summarizes the number of articles in our previous study and current research. Overall, if we look at the average number of articles mentioning transplantation that were published in a month, we can see that their number increases in time, which reflects a gradual increase in the interest of both the editors and the readership.

3 The initial search resulted in 167 articles, out of which ten were removed due to the lack of relevance for the topic.

Table 1. Total number and percentage of articles on organ donation in the selected periodicals (2008–2015)

Current number	Year	Periodical name	Total number of articles on organ donation	Percentage (number of articles in one publication/month)
1	2008*	*Adevărul*	67	5.58
2	2008*	*Cotidianul*	43	3.58
3	2010**	Quality press (3 titles, including *Adevărul*)	124	3.44
4	2010**	Tabloid press (3 titles, including *Click!*)	139	3.86
5	2015 ***	*Adevărul*	85	7.08
6	2015 ***	*Click!*	72	6.00

* Data for 2008 is retrieved from Frunză et al. 2009; ** Data for 2010 is retrieved from Holman et al. 2012. *** Data from 2015 has not been published before.

In our analysis of the articles from 2015, we have followed a qualitative rather than a quantitative perspective; thus, we were guided by what Moshe Idel labels "an eclectic methodology". When he performs the analysis of religious phenomena, he follows the "major concerns that define the specificity of particular styles" (Idel 2007). Similarly, we were interested in discursive patterns and strategies that emerged both from a diachronic reading of the articles and from a synchronic reading of the whole textual corpus.

When discussing the main topics covered by the two newspapers, the articles from both *Adevărul* and *Click!* display a variety of transplanted organs and tissues (head, hands, heart, kidney, liver, pancreas, small bowel, lungs, penis, hair, skin, cornea, ovary, and bone marrow), although only a minority of transplant procedures are described as being actually performed in the Romanian context (kidney, liver, heart, skin, cornea, or bone marrow).

There were three major categories of articles in both newspapers that emerged after the analysis: 1) articles depicting cases of transplants performed/planned to be performed in Romanian transplant centers (35% of all articles); 2) articles describing/diagnosing the general situation of transplantation in Romania, without focusing on a particular case (18% of all articles); and 3) articles depicting cases of transplants performed/planned to be performed abroad (42% of all articles). A small percentage of articles (5%) had topics that did not fit into the three categories and were too unrelated to fit into a new one. Within the first category, a subcategory found massively in *Click!* consisted of articles depicting a Romanian celebrity who either had undergone a transplant procedure a while ago (in Romania) or has been waiting for a transplant during the time frame of the article (40% of all articles from *Click!*).

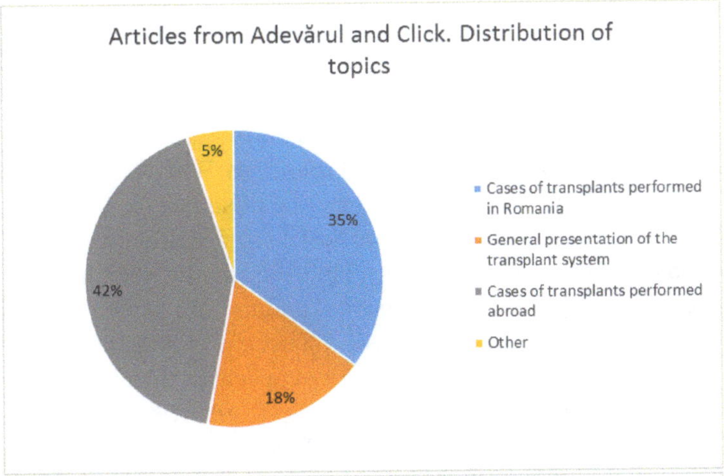
Articles from Adevărul and Click. Distribution of topics
5%
35%
42%
18%
Cases of transplants performed in Romania
General presentation of the transplant system
Cases of transplants performed abroad
Other

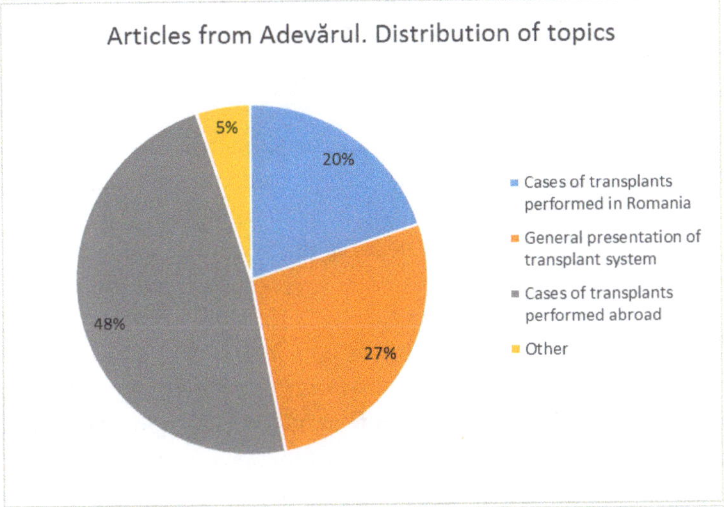
Articles from Adevărul. Distribution of topics
5%
20%
48%
27%
Cases of transplants performed in Romania
General presentation of transplant system
Cases of transplants performed abroad
Other

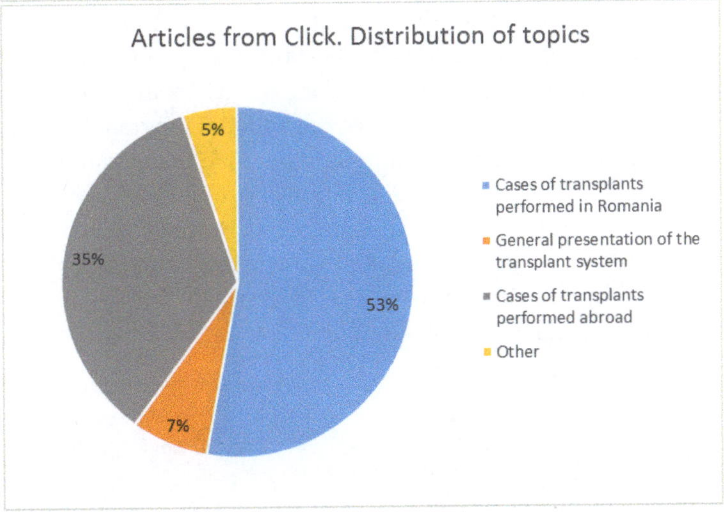
Articles from Click. Distribution of topics
5%
35%
53%
7%
Cases of transplants performed in Romania
General presentation of the transplant system
Cases of transplants performed abroad
Other

The articles from the first category (transplants performed/expected in Romania) generally display a duality of perspectives: that of the happy recipient who has been given a new life chance, and that of the donor for whom nothing could be done from a medical perspective. When doctors are being cited in these articles, they speak about complications, problems, and financial limitations. When recipients or donor relatives are cited, they express a variety of emotions (happiness, grief, and concerns) (Bochiș 2015; Ion 2015). The subcategory "celebrity articles" from *Click!* usually revolves around a celebrity who is displayed in vivid and detailed pictures, replete with sensational details about former lovers, scandals, and personal (mis) fortune. The issue of transplantation is mentioned somewhere as accessory, even if it represents the main element why the said celebrity appears in the newspaper. Moreover, 14% of articles from *Click!* describe the case of a celebrity (whose daughter also works in the showbiz field), who tragically died of kidney failure (transplant was scheduled but the procedure could not be performed because of medical complications); however, the main issue covered in the article was the disputed mother–daughter relation with all the fights, insults, and obscenities exchanged between them during the mother's lifetime[4].

The second category of articles displays various presentations of the overall situation of the transplant system in Romania (citing statistics, comparing the relative performance of local centers). Regional discrepancies are particularly commented upon: thus, "Oradea occupies the first place among centres, after Bucharest" (Bonchiș 2015), or "Bucharest has been overpassed, now the donors from Ardeal are more numerous than in the capital city" (Spiridon 2015), or "Half of the kidneys from Moldova 'fly' to the capital city" (Ciuhu 2015). We find that there are many centers with little-to-no transplant activity: only one donation performed in an entire year in Sălaj (Pop 2015) or in Vîlcea (Rîpan 2015). In addition, there are announcements for inauguration of new centers or programs, which are not yet functional: the most notable example is the new transplant center in Brașov, which is announced as "almost" functional over the entire year 2015; various financial problems prevent its opening (six articles in *Adevărul*: Suciu 2015a, 2015b, 2015c; Dan 2015a, 2015b, 2015c). Other examples are the center for pediatric heart transplant in Bucharest, to be opened some time during 2015; or the national program for lung transplantation, which is still in preparation, although there are patients in need. This category of articles is underrepresented in *Click!* (only 7% of articles).

The third category of articles describes transplants performed or planned abroad. One subgroup depicts world premieres (penis transplant, hand transplant, ovary transplant, head transplant, and so on) (Ștefan 2015a, 2015b; Băltărețu 2015). The other subgroup covers cases of Romanian patients who are either

4 Among these articles, one can see, for instance, those by Albert 2015; Văcaru 2015; Pavel 2015.

unable to find their treatment in their own country or are actively seeking to find a transplant procedure abroad (for which they attempt to raise funds). Here, one encounters cases of desperate people who are depicted as being left at the mercy of their fellows: parents attempting to raise money for their children; students raising money for their colleagues; patients contemplating risky and controversial procedures abroad. As most of these cases are merely indicated and not followed afterward, readers may wonder whether people's pain and suffering are adequately depicted in these articles.

Another finding is that the image depicted in the media is profoundly disconnected with the pressing issues of organ donation. If one asks the professionals about the most pressing problems that the transplant system is facing, one will usually find out about the great discrepancy between the number of available organs and the long waiting list; the lack of education of the general public on matters of organ donation; and the relevance of the consent type for increasing donation rates (Frunză and Gavriluță 2012). However, when reading the newspapers, among the most pressing issues in transplant for the Romanian society seem to be head transplant (not less than 8% of articles from both *Adevărul* and *Click!*) (Ștefan 2015a; Băltărețu 2015) or (perhaps surprisingly) penis transplantation (Ștefan 2015b, 2015c). These very unlikely (at least for Romania) types of transplant are discussed in great detail with lots of visual elements, in both the tabloid periodical and the quality one.

Unsurprisingly, there are numerous ethical issues that might be discussed in relation to these articles, when we approach the texts from an ethical standpoint. One worrying finding is that we found numerous cases of incomplete articles that provide misleading, if not outright false, information. Unfortunately, we are not referring to inexperienced media outlets and dubious websites, but of two well-known and widely read publications with large editorial teams and staff teams with licensed journalists. One telling example was an article on penis transplantation entitled: "Good news: penis transplantation in Romania." Closer reading of the body of the article revealed that this type of transplant is not yet performed in Romania and remains a distant possibility in the future (Ștefan 2015b).

We can accept that these articles can raise awareness of the general audience about the delicate and complex aspects of human existence that organ donation entails. We can also safely assume from the comments that at least some readers come across the transplant articles only because of its presentation as a taboo story for the public debate and the novel approach that goes beyond the journalist ethical code. But apart from this possible positive effect of increasing topic awareness, we believe the incorrect depiction of such a delicate and controversial issue will only hurt efforts to inform the audience and shape their donation behaviors. It is difficult to assume that a tabloid depiction of a sensitive issue can lead to a valuable input at the level of moral decision-making.

Ethical requirements for public communication (Frunză and Frunză 2011) should be reinforced when dealing with "vulnerable populations", a category that traditionally encloses children, incapacitated individuals, and patients in general, due to their level of dependence on the medical system (Frunză et al. 2016; Grad 2015; Sandu and Caras 2013; Loue 2000). Furthermore, we need to pay attention to deontological aspects when influencing and commercializing people's suffering and pain. This is especially appalling when the subjects of the suffering are small children. As such, one of the cases of hand transplantation is illustrated with the picture of a youngster and his mother, displaying the moving headline: "Look, mom, I've got hands! First transplant to the youngest patient" (Damian 2015).

One can further note the unnecessary repetition of subjects and topics, even if the public interest is minimal. The same stories are rerun many times, even though the level of public involvement (assessed by comments, social media interaction) is essentially non-existent. For instance, the head transplant announced as novelty for 2017 and projected to be performed by Dr. Canavero on a wealthy Russian patient has been extensively covered (including videos), even though the event was only a distant possibility during the time frame of 2015 (Ștefan 2015; Băltărețu 2015).

Sadly, there are still some cases of plagiarized articles (pieces copy/pasted from science magazines). The practice of plagiarism and this omission of sources in the media, apart from the practice of copy/paste, are ethically significant in the Romanian cultural context, wherein accusations of plagiarism are endemic in the media debate, especially in association with the scientific papers of widely known public actors (Șercan 2017). Certainly, this issue is of utmost importance due to the growing ethical responsibilities that the journalism profession implies (Frunză and Frunză 2011).

Concerning the topic of deceased donation, one can note that a small number of articles reveal the identities of donors by including pictures and recognizable personal details, even if the law requires maintaining their anonymity. (Ion 2015; Rotaru 2015; Bonchiș 2015; Both 2015). Even more seriously, there is the issue of disputable usage of minors' photos (whose pictures would have to be blurred or made unrecognizable) (Rotaru 2015).

All these ethical problems are endemic not only for transplant stories, but in general for the wider Romanian media. However, if tabloidization is a symptom, the consequences of a bad story, or a poorly told story, can be extremely harmful for the medical field in general and for the transplant field in particular, where the issue of trust plays such an important role (Frunză et al. 2012). When we analyze the content of the two newspapers, the marks of tabloidization are apparent, both in the tabloid paper and in the quality one: articles with deceiving headlines, with unreliable content, with redundant repetition of the same topics, and articles breaking the deontological code of journalists. The proportion of deceiving articles is higher in *Click!* (75%) than in *Adevărul* (25%), but the trend of tabloidization is present in both newspapers.

7.4 From mutual ignorance to mutual partnerships: students' campaigns promoting organ donation

In this part, we want to change the tone and evaluate some successful (even if limited) public communication campaigns enacted by students. We do not want to imply that students' contributions are enough or that they compensate for the lack of a planned public campaign. However, we do wish to emphasize that peer-to-peer promotion campaigns in favor of organ donation, such as these, represent a successful trend in the field of communication campaigns on this topic.

Researchers investigating this topic emphasize the enthusiasm and professionalism brought by the so-called "digital natives" when performing campaigns, their good knowledge of the habits of their peers, and the efficiency of combining education, information, and entertainment.

We provide the example of two student campaigns promoting organ donation, based in Cluj and analyzed in detail elsewhere (Frunză and Guga 2017). Both campaigns were developed by students from the Faculty of Political, Administrative, and Communication Science at Babeș-Bolyai University as a course-based request (this did require the format of a promotion campaign but did not impose the content of the campaign, i.e., students freely chose donation campaigns).

The two campaigns performed by students managed to combine a variety of activities; for instance:

- mobilize local celebrities (football players) to endorse the campaign;
- organize formal educational events (workshops) with experts;
- organize nonformal educational events (a treasure hunt with clues);
- mobilize online communities (via Facebook pages, Facebook events, personal Facebook accounts).

Both campaigns managed to attract the attention of traditional media, and their messages were disseminated in the local press. In both articles, the content was positive and reflected the message of the campaign.

The potential weaknesses from an ethical perspective concern the accuracy of the message. Additionally, these campaigns depend on the enthusiasm of their initiators, which cannot be prescribed. If we take into account the other student peer-to-peer campaigns from the UK and the US, we can affirm that these campaigns manage to construct partnerships with the media and, thus, improve the way audiences are informed on these topics.

7.5 Conclusions

The encouragement of a positive public attitude toward organ transplantation and cultivation of public support for this medical procedure is essential for a functional transplant system. Media have the capacity to influence the public attitudes to health-related behaviors. In Romania, in the absence of a coherent public campaign on organ donation, media remains the major source of information on organ transplantation.

Referring to previous Romanian studies on the topic published, this article underlines the growing interest manifested in Romanian media for the topic of organ transplantation, as suggested by the increasing number of articles on the subject. However, the analysis of the content of the articles shows that what the Romanian media presents to the public is many times (75% in *Click!*, 25% in *Adevărul*) a "tabloid" picture of the organ transplantation process, characterized by quasi-relevance of the content to the topic of organ transplantation, neglecting of the actual pressing issues related to organ transplantation, and ethically flagrant practices that have a negative impact on the field of transplantation as it is perceived by the Romanian readership. Yet, by analyzing the phenomenon of peer-to-peer campaigns promoting organ donation, the article draws attention to the resources offered by both traditional and new media to promote organ transplantation.[5]

References

Abadie, Alberto, and Sebastien Gay. 2006. "The impact of presumed consent legislation on cadaveric organ donation: a cross-country study." *Journal of health economics* 25.4 (2006): 599-620.

Albert, Alexandra. 2015. "Ultimele date despre Marioara Zăvoranu. Medicii vor transplant renal." *Click!*, January 28, 2015. http://www.click.ro/vedete/romanesti/ultimele-date-despre-marioara-zavoranu-medicii-vor-transplant-renal

Băltărețu, Răzvan. 2015. "Primul transplant de cap din lume şi-a găsit voluntar: cine este rusul dispus să-şi doneze capul." *Adevărul*, April 16 2015, 12:45, http://adevarul.ro/sanatate/medicina/primul-transplant-cap-lume-si-a-gasit-voluntar-rusul-dispus-sa-si-doneze-capul-1_552f8284cfbe376e352b29cf/index.html

Dan, Sebastian. 2015a. "Dacă nu apar contestaţii, Centrul de transplant renal intră în funcţiune la sfârşitul lunii iulie." *Adevărul*. June 4, 2015. http://adevarul.ro/locale/brasov/daca-nu-apar-contestatii-centrul-transplant-renal-intra-functiune-sfarsitul-lunii-iulie-1_557020efcfbe376e35e97224/index.html

Dan, Sebastian. 2015b. "Fără contestaţii la licitaţii, Centrul de Transplant Renal va fi dotat cu aparatură în octombrie." *Adevărul*. September 4, 2015. http://adevarul.ro/locale/brasov/fara-contestatii-licitatii-centrul-transplant-renal-dotat-aparatura-octombrie-1_55e93a6cf5eaafab2c3d6576/index.html

Damian, Andreea. 2015. "Mami, uite acum am şi eu mâini! Transplant în premieră pentru cel mai tânăr pacient." *Click!*, July 29, 2015 - 21:52, http://www.click.ro/news/lume/mami-uite-acum-am-si-eu-maini-transplant-premiera-pentru-cel-mai-tanar-pacient

5 Mihaela Frunză acknowledges the financial support from the Swiss National Science Foundation for participating in the Bucharest Bioethics Symposium (2016) and for writing this paper. She also wishes to acknowledge the constant support received from the ELPAT (www.elpat.org) over the past years, especially (for this article) from the working group Public Opinion.

Dan, Sebastian. 2015c. "S-a finalizat licitaţia pentru aparatura centrului de transplant renal." *Adevărul*. November 2, 2015. http://adevarul.ro/locale/brasov/s-a-finalizat-licitatia-aparatura-centrului-transplant-renal-1_5636f909f5eaafab2c3b75cb/index.html

Feeley, Thomas Hugh, and Shin-Il Moon. 2009. "A meta-analytic review of communication campaigns to promote organ donation." *Communication Reports* 22.2: 63-73

Frunză, Mihaela, Sandu Frunză, Cătălin Vasile Bobb, and Ovidiu Grad. 2010. "Altruistic living unrelated organ donation at the crossroads of ethics and religion. A case study." *Journal for the Study of Religions and Ideologies*, vol. 9, nr. 27, Winter 2010: 3-24.

Frunza, Mihaela-Cornelia, Ioana Iancu, Sandu Frunza, Iulia Grad, Delia Balaban, and Ovidiu Grad. 2011. "Media Analysis of Romanian Debates on Presumed Consent – Ethical and Legal Aspects" *Organ Transplantation: Ethical, Legal and Psychosocial Aspects, Vol. II Expanding the European Platform*, Eds. W. Weimar, M. A. Bos, J. J. Busschbach. Pabst Science Publishers, Lengerich. 218-228.

Frunză, Sandu and Mihaela Frunză. 2011. "The Ethical Fundamentals of the Responsibility of the Journalists." *Journal of Media Research*, vol. 4 issue 1(9)/2011, 33-43.

Frunză, Mihaela and Ariana Guga. 2017. "Partnerships between media and ethics: Students peer-to-peer campaigns for organ donation." In *Ethical, Legal and Psychosocial Aspects of Transplantation*, Vol. IV. Global Challenges, Ed. W. Weimar, Pabst Science Publishers, Lengerich. 213-221.

Frunză, Sandu. 2011. "On the Need for a Model of Social Responsibility and Public Action as an Ethical Base for Adequate, Ethical and Efficient Resource Allocation in the Public Health System of Romania." *Revista de cercetare şi intervenţie socială*. vol. 33: 178-196.

Frunză, Mihaela, Iulia Grad, and Sandu Frunză. 2016. "Ethical Aspects of Gender Stereotypes in Romanian Advertising." *ESSACHESS. Journal for Communication Studies*, vol. 9, no. 1 (17): 143-157;

Gimbel, Ronald W., Martin A. Strosberg, Susan E. Lehrman, Eugenijus Gefenas, and Frank Taft. 2003. "Presumed consent and other predictors of cadaveric organ donation in Europe." *Progress in Transplantation* 13.1 (2003): 17-23

Grad, Iulia. 2015. "Ethical Perspectives on Mediated Communication." *Journal for the Study of Religions and Ideologies*, vol. 14, issue 41 (Summer 2015): 224-242

Holman, Andrei, Adina Karner-Huţuleac, Cristina Gavriluţă, Mihaela Frunză, Irina Streba, Lacrima Boilă, and Beatrice Ioan. 2012. "Themes and semantic patterns of the Romanian Press portrayal of organ transplantation." Paper presented at the *11th World Congress of Bioethics*, Rotterdam, the Netherlands, 26-29 June 2012.

Holman, Andrei, Adina Karner-Huţuleac, and Beatrice Ioan. 2013. "Factors of the willingness to consent to the donation of a deceased family member's organs among the Romanian urban population." *Transplantation Proceedings* Vol. 45, No. 9: 3178-3182.

Idel, Moshe. 2010. "Abordări metodologice în studiile religioase/Methodological Approaches in Religious Studies." *Journal for the Study of Religions and Ideologies*, 6(16), 5-20.

Ioan, Beatrice, Cristina Gavriluta, Andrei Holman, Mihaela Frunza, Irina Streba, Adina Karner-Hutuleac, Roxana Zavoi, and Lacrima Boila. 2011. "Level and Sources of Information on Organ Transplantation in Romania." *Transplant International*, vol. 24, Suppl. 2, (September 2011): 325-326.

Jung, Harald and Mihaela Frunza. 2014. "Attitudes towards living organ donation in two Romanian universities." In *Ethical, Legal, and Psychosocial Aspects of Transplantation. Global Issues, Local Solutions,* eds. Willem Weimar, M.A. Bos, J.J. Busschbach. Pabst Science Publishers, Lengerich.

Kaiser, Gernot Maximil, Sonia Radünz, Sidney Becker, Sumeska Thavarajah, Andreas Paul, and Matthias Heuer. 2014. "Organ donation: new hope through the expected amendment in Germany?" In *Transplantation proceedings* Jul-Aug;46(6):2040-2.

Karner-Huțuleac, Adina. 2012. "The Discourse Analyse in the Romanian Media Regarding the Organ Transplant–Systemic References." *Revista Românească pentru Educație Multidimensională*, 129.

Krekula, Linda G., Silvia Malenicka, Margareta Linder and Annika Tibell. 2009. "From words to action–influence of two organ donation campaigns on knowledge and formal decision making." *Clinical transplantation*, *23*(3), 343-350.

Loue, Sana. 2000. *Textbook of research ethics: Theory and practice*. Springer Science & Business Media.

Mossialos, Elias, Joab Costa-Font, and Caroline Rudisill. 2008. "Does organ donation legislation affect individuals' willingness to donate their own or their relative's organs? Evidence from European Union survey data." *BMC health services research* 8.1 (2008): 48

Morgan, Susan E. and Tyler R. Harrison. 2010. "The Impact of Health Communication Research on Organ Donation Outcomes in the United States." *Health Communication*, 25:6-7, 589-592.

Pavel, Ana. 2015. "Marioara Zăvoranu a murit. De ce boli suferea." *Click!*, April 15, 2015. http://www.click.ro/vedete/romanesti/marioara-zavoranu-murit-de-ce-boli-suferea

Pop, Alina. 2015. "O polițistă din Sălaj a hotărât să devină, post-mortem, donator de organe: "Acolo unde mergem cu toții, după moarte, nu avem nevoie de organe interne". *Adevărul*, January 7, 2015. http://adevarul.ro/locale/zalau/o-politista-salaj-hotarat-devina-post-mortem-donator-organe-acolo-mergem-totii-moarte-nu-nevoie-organe-interne-1_54acc7b9448e03c0fd5f1b36/index.html

Rithalia, Amber, Catriona McDaid, Sara Suekarran, Lindsey Myers and Amanda Sowden. 2009. "Impact of presumed consent for organ donation on donation rates: a systematic review." *British Medical Journal, 338*, a3162.

Rîpan, Irina. 2015. "Drama unui tânăr abandonat într-un orfelinat care nu și-a cunoscut niciodată mama: dependent de dializă, are 22 de ani, dar arată de 12 ani." *Adevărul*, June 5, 2015. http://adevarul.ro/locale/ramnicu-valcea/drama-unui-tanar-abandonat-intr-un-orfelinat-nu-si-a-cunoscut-niciodata-mama-dependent-dializa-22-ani-arata-12-ani-1_55719ab0cfbe376e35f5875b/index.html

Rodríguez-Arias, David. 2013. "The social acceptability of implementing morally challenging strategies to increase donor rates." *Public engagement in organ donation and transplantation*, ed. G. Randhawa and S. Schicktanz, 70-79.

Sandu, Antonio and Ana Caras. 2013. "Deconstruction of Charity. Postmodern ethical approaches." *Journal for the Study of Religions and Ideologies*, vol. 12, issue 36 (Winter 2013): 72-99.

Schwettmann, Lars. 2015. "Decision solution, data manipulation and trust: The (un-) willingness to donate organs in Germany in critical times." *Health Policy*, *119*(7), 980-989.

Suciu, Simona. 2015a. "Centru de transplant renal la Brașov din acest an." *Adevărul*. January 25, 2015, http://adevarul.ro/locale/brasov/centru-transplant-renal-brasov-an-1_54c6274d448e03c0fdf98e54/index.html.

Suciu, Simona. 2015b. "Centru de Transplant Renal, din luna mai la Brașov." *Adevărul*. February 27, 2015.: http://adevarul.ro/locale/brasov/centru-transplant-renal-lunamai-brasov-1_54f015a2448e03c0fd07ed3d/index.html

Suciu, Simona. 2015c. "Consiliul Județean a dat banii pentru dotarea Centrul de Transplant Renal. Ministerul Sănătății a făcut doar promisiuni." *Adevărul*, March 27, 2015. http://adevarul.ro/locale/brasov/consiliul-judetean-dat-banii-dotarea-centrulde-transplant-renal-ministerul-sanatatii-facut-doar-promisiuni-1_5514ed6a448e03c0fdf37fb5/index.html

Șercan, Emilia. 2017. *Fabrica de doctorate sau Cum se surpă fundamentele unei nații*, București: Humanitas.

Ștefan, Carmen. 2015a. "Bolnavii își vor putea transplanta capul pe alte corpuri sănătoase din 2017" VIDEO, *Click!*, February 27,2015 – 12:34, http://www.click.ro/news/lume/bolnavii-isi-vor-putea-transplanta-capul-pe-alte-corpuri-sanatoase-din-2017-video

Ştefan, Carmen. 2015b. "Veste bună: transplant de penis în România." *Click!*, July 28, 2015. http://www.click.ro/news/lume/veste-buna-transplant-de-penis-romania

Ştefan, Carmen. 2015c. "Va primi un penis nou după ce a fost rănit grav în Afganistan." *Click!*, December 9, 2015. http://www.click.ro/news/lume/va-primi-un-penis-nou-dupa-ce-fost-ranit-grav-afganistan

Văcaru, Clara. 2015. "Reacţia crudă a Oanei Zăvoranu, după ce şi-a văzut mama pe patul de spital." *Click!*, April 8, 2015. http://www.click.ro/vedete/romanesti/reactia-cruda-oanei-zavoranu-dupa-ce-si-vazut-mama-pe-patul-de-spital

Van Dalen, Hendrik P., and Kene Henkens. 2014. "Comparing the effects of defaults in organ donation systems." *Social science & medicine*, *106*, 137-142.

Wakefield, Melanie A., Barbara Loken and Robert C. Hornik. 2010. "Use of mass media campaigns to change health behavior." *Lancet*, 2010 376(9748): 1261–1271.

Mihaela Constantinescu

8 Seeing the Forest Beyond the Trees: A Holistic Approach to Health-Care Organizational Ethics

8.1 Introduction

With the growing body of research concerning organizational ethics of health care over the past 2 decades (Barina 2014; Gallagher and Goodstein 2002; Hall 2000; Khushf 1998; Magill and Prybil 2004; Potter 1996; Rorty et al. 2004; Werhane 2000), ethicists have turned their attention to the way in which organizational contexts have a direct influence on the practice of health care (Bean 2011; Førde and Hansen 2014; Khushf 1998; Spencer et al. 2000). Against the background of continuous transformations in the way health-care organizations (HCOs) operate, contextual aspects, such as financial pressure, time efficiency, or reporting to supervisory bodies, are often underlined as influencing medical decisions made by physicians and staff (Austin 2007; Carney 2011; Wesorick 2002), sometimes to the detriment of patients' needs (Hart 2005). Ongoing moral concerns over institutional processes may potentially lead to structural tensions and conflicts among staff (Barina 2014; Gibson 2012), or even inappropriate medical outcomes (Chen et al. 2007), suggesting that there is still room to further develop organizational health-care ethics.

Despite the focus on organizational level of moral issues, contextual features of HCOs are still approached in a dichotomous manner. One prominent example concerns compliance versus integrity, often understood as a rules *or* values approach to organizational ethics and translated into conflicting and disconnected ethics programs within HCO practice (Boyle et al. 2000; Magill and Prybil 2004; Mills and Spencer 2001; Silverman 2000). This could negatively influence future development of the field, leading to fragmentation and lack of consistency. I argue that, if we want to advance research on organizational ethics and gain more explanatory power, we need to approach such dichotomies holistically, in a manner that is integrative instead of antithetical. To that end, I argue that compliance and integrity should be seen as complementary approaches to ethical issues occurring at the organizational level and that they respond to and mirror two organizational dimensions that require better alignment: the organizational informal culture and formal structure. Both are needed to make ethics work in HCOs[1].

---

1 Throughout this paper, by health-care organizations (HCOs), I refer to "medium- and large-sized provider organizations that have a defined management structure" (Werhane 2000, 170) and whose focus is on maximizing the healing and well-being of patient population (Wall 2007; Werhane 2000).

To address this issue, I use normative insights from research in business ethics. First, I make a brief overview of organizational ethics in health care and point out its analogous counterpart in business ethics – ethics management. Second, I argue against the common dichotomy between integrity and compliance in HCOs, by referring to a normative model of integrated ethics management advanced in business ethics research. Third, I explore how an optimal alignment between organizational culture and structure can better embed ethics at an organizational level in health-care settings. I end with suggestions for future research, highlighting the need for closer collaboration between research in HCO ethics and business ethics.

8.2 Organizational ethics in health care

Since its emergence as a field in the mid-1990s, organizational ethics in health care has been interpreted in various ways, all of them anchored in the idea that HCOs *per se* are responsible for the outcomes of their operations at a distinct level from their constitutive members (Spencer et al. 2000). Research on organizational ethics therefore acknowledges that the medical act is influenced by the way work is framed in the health-care setting (Carney 2011; Fox et al. 2012; Rorty et al. 2004). This is especially relevant when the ethical quality of clinical care is at stake (Førde and Hansen, 2014), given that practitioners often cannot control the policies and procedures of their respective institutions (Emanuel 1995; Gallagher and Goodstein 2002; Reiser 1994).

As a result, organizational ethics, unlike bioethics or clinical ethics that deal with specific ethical issues occurring in medical practice, is concerned with "all aspects of the operation of the HCO so that positive ethical climate can be developed and maintained" (Spencer et al. 2000, 5). It consists of various "processes to address ethical issues associated with the business, financial, and management areas of health care organizations (HCOs), as well as with professional, educational, and contractual relationships affecting the operation of the HCO" (Spencer et al. 2000, 212). As such, organizational ethics refers to the overarching framework by which processes, procedures, and policies are designed to ensure that organizational performance is in line with its ethical foundations (Phillips and Margolis 1999). To that end, HCOs are required to work toward defining key values, while also striking an optimal balance between conflicting values and stakeholder expectations.

Briefly put, three definitional aspects may be delineated to elucidate the meaning of organizational ethics: (1) issues pertaining to management and governance of HCOs; (2) implications of organizational decisions on stakeholders (practitioners, staff, patients, regulatory bodies, the industry, and so on); and (3) complexities of balancing key organizational goals, such as quality of patient care, financial sustainability, staff management, medical research, and public accountability (Gibson 2012; Gibson et al. 2000; Hall 2000; Spencer et al. 2000).

This understanding of HCO ethics is synonymous to the notion of "ethics management" used in business ethics. Kaptein (1998, 42) defines ethics management as "the systematic and coherent development of activities and the taking of measures in order to realize the fundamental and justified expectations of stakeholders and to balance conflicting expectations of stakeholders in an adequate way". For Jeurissen[2] (2004, 11), ethics management is centered on the question "how do you manage ethics in organizations?" and "aims at improving the decision-making processes, the procedures and structures in an organization, so that the operations of the organization are more geared towards ethical principles". To that end, organizations use instruments such as codes of ethics, ethical audits, ethical trainings, or ethics hotline.

The focus on ethical issues at the organizational level has marked a shift in business ethics from a bad apples to a bad barrels paradigm (Treviño and Youngblood, 1990), in a move forward that has resulted in more emphasis being put on "the characteristics of the organizational context within which unethical behaviour occurs" (Kaptein 2011, 844) than on "the personal characteristics of individual transgressors" (Idem). *Mutatis mutandis*, we need to acknowledge that errors in HCOs often reflect faulty systems rather than practitioner errors (Austin 2007) and that HCOs need "an ethics *of* the system rather than ethics *in* the system" (Wall 2007, 228). However, the question remains as to how should ethical issues pertaining to the HCO context be best approached?

8.3 Compliance versus integrity: an apparent dichotomy

Research about managing ethics at the organizational level in health care proposes one apparent dichotomy between compliance and integrity (Mills and Spencer 2001; Silverman 2000), with organizational ethics being equated with the latter. Overall, a rules-centered or compliance approach is interpreted as focusing only on legal, regulatory, or administrative norms, whether externally or internally imposed, and on how to effectively enforce these guidelines. Instead, a values-centered or integrity-based approach is seen as focused on educating and guiding ethical judgments and behavior of organization members, usually by rewarding excellence (Boyle et al. 2000; Magill and Prybil 2004; Weaver and Treviño 1999).

2 In the business ethics context, Jeurissen uses the notion of organizational ethics to refer to the type of applied ethics aimed at "analysing specific ethical problem types in organizations, in order to provide normative clarification and guidance" (2004, 11), with examples such as advertising ethics, the ethics of insider trading, or the ethics of company restructuring. Translated to health care, this characterization is rather synonymous to clinical ethics. Therefore, I take common understanding of organizational ethics in health care to be rather synonymous to what business ethicists call "ethics management".

This theoretical dichotomy was supported in health-care practice by the way the two types of programs were introduced at the same time in the USA (Mills and Spencer 2001). On the one hand, government agencies such as the Department of Justice (DOJ) focused on reducing fraud and abuse in health care, imposing specific compliance programs that HCOs could easily implement to meet the specific requirements of the Federal Sentencing Guidelines for Organizations. On the other hand, the Joint Commission for Accreditation of Healthcare Organizations (JCAHO) started, in 1996, to focus on the effect of organizational context on medical performance. To meet accreditation requirements, HCOs needed to develop a process that integrates ethical issues pertaining to various departments and protects the integrity of clinical decision-making, a process they named organizational ethics.

In practice, this resulted in organizations implementing compliance programs on a large scale, but with implementation of fewer integrity programs (Mills and Spencer 2001). Sometimes, compliance programs are implemented as a substitute for, or even a competitor to, integrity or organizational ethics programs (Boyle et al. 2000). In other cases, HCOs have developed two compartmentalized programs (Mills and Spencer 2001), one targeted toward government compliance and the other involving more diffuse ethics activities. While compliance programs emphasize interdictions, integrity programs highlight what should be done in order to strive for moral excellence (Boyle et al. 2000; Silverman 2000). In this context, some have already indicated that organizational ethics strategy must go beyond compliance programs (Magill and Prybil 2004), while others point to an integration of the two (Mills and Spencer 2001) within the existing infrastructure of compliance programs. However, there is still a paucity of normative research concerning the integration of compliance and integrity in HCO ethics.

Developments in normative grounding may be found in the business ethics interpretation of ethics management, where the polarization between rules and values used to be the dominant interpretative view (Jeurissen 2004). However, business ethics has evolved to a more refined approach, in which an integrative view is favored against the former antithetic approach. Organizational ethics in health care could adopt this holistic approach, leading to a more comprehensive perspective of compliance and integrity issues at the organizational level.

Instead of being opposites, these two approaches differ only by their *degree of moral complexity* (Jeurissen 2004). Issues pertaining to compliance and integrity are placed on a continuum of increased moral complexity, without being mutually exclusive. Namely, Jeurissen (2004) posits rules-based and values-based approaches as the first two stages of moral development in an integrated four-level model of ethics management. Depending on the moral complexity of the situation, we may rely on one possible approach. This largely depends on two main coordinates of the situation: the action context and the normative context. While the first refers to the complexity of the issue under moral evaluation, the latter pertains to the complexity of the overall stakeholder framework. Both the action and the normative context involve a low and a high level of complexity: the higher the complexity, the broader the instruments of ethics management that need to be used to address increasingly diverse stakeholder groups. This combination finally results in four approaches or levels of ethics management that respond to the increase of situational moral complexity in organizations.

The four levels move from simple to complex. First is a rules-based approach, adequate when basic moral criteria need to be applied to standard cases. Second involves a values-based approach, relevant when creative solutions are needed for new ethical problems. Third comprises a stakeholder dialogue approach, required when there is moral disagreement among stakeholders. Fourth and last is a social dialogue approach for occasions when moral disagreement involves larger societal spheres and issues. As Jeurissen (2004) puts it, these levels are caught in a contingent and evolutionary relationship. This means that while all four approaches reinforce and sustain one another, each retains its own unique relevance to ethical matters in organizations. On the other hand, this also means that they evolve from simple to complex issues, but at the same time, none is "better" than its predecessor because in practice, each of these approaches "can be an adequate solution to specific types of ethical problems, depending of situational characteristics" (Jeurissen 2004, 16).

The model may be easily translated to health care. HCOs display several characteristics that differentiate them from other types of organizations (Werhane 2000). Boyle et al. (2000, 10) identify four such characteristics: (1) their mission to alleviate pain and suffering and restore health; (2) the complex, highly regulated environment – internal and external – in which they operate; (3) professional cultures (physicians, nurses, health care managers); and (4) the rapidly changing health care market. Based on these characteristics, we could say that HCOs operate under all four levels of moral development as identified by Jeurissen (2004). First, a HCO includes multiple routinized activities, which allow it to "function efficiently and effectively and also enable its stakeholders to understand it and rely on it" (Wall 2007), given the predictability of the medical protocols. Routine organizational situations should be captured by a rules-based strategy, efficient when clear and unequivocal guidelines are needed (Jeurissen 2004). Instead, more complex issues require a value-based strategy, for instance, aspects such as providing care to the uninsured, research funding for new treatments, or staffing policies. Other situations require HCOs to address moral issues, as well as societal concerns (e.g., new medicine testing or genetic modification), thus yielding an even higher level of complexity.

This model provides some relevant normative guidance to dissolve the dichotomy between compliance and integrity in HCO ethics. The next question is this: what does this mean for how organizations operate?

8.4 Embedding ethics in HCOs: optimal alignment between culture and structure

Once we overcome the apparent antithesis between rules and values, and capture the broader picture of organizational ethics, we can explore how this may be reflected in the way HCOs function. With both compliance and integrity sharing a common goal, namely, to improve the ethical climate and relationship with key stakeholders, we need to research how they may be integrated in the organizational context.

To achieve organizational integrity, one has to combine two operational components: the informal culture – values, expectations, and unwritten norms – and the formal structure – processes, procedures, and rules (Constantinescu and Kaptein 2015b; Kaptein and Wempe 2002; Silverman 2000). Both form the overall context in which the medical act takes place and together build up the organizational practices that "are *actually* expressed in the actions of organizational members" (Kaptein and Wempe 2002, 146–9). Therefore, organizational ethics is focused on the actual behavior occurring in health-care settings as a result of the organizational context (Magill and Prybil 2004).

The complementary relationship between compliance and integrity is reflected in the relationship between the formal structure and informal culture. While compliance is linked to the organization's formal structure, integrity is more useful for dealing with aspects of the informal culture. The formal structure determines what constitutes ethical behavior (Kaptein 2008), while the informal culture stimulates ethical conduct (Weaver and Treviño 1999). There is a certain reciprocity between organizational ethical principles and ethical conduct: ethical principles adopted at the organizational level inspire ethical behavior, and ethical behavior reinforces the organizational ethical principles (Magill and Prybil 2004). This could potentially lead to a mutually enhancing relationship between an individual's ethical behavior and organizational ethical practices. The more the individuals behave ethically, the more the organizational practices become ethical, and in turn, stimulate individuals to act even more ethically. Individual and organizational practices build off each other in a cycle, analogous to an upward double-helix relationship (Constantinescu and Kaptein 2015a).

Given this interdependence, research on both business ethics and organizational ethics highlights the need to have consistent formal policies and informal culture within an organization, as well as to balance "what is formally expected and the ways things really get done" (Boyle et al. 2000). For formal procedures and policies to be effective, they need to be linked to the informal organizational culture (Chen et al. 2007; Treviño et al. 1999; Weaver and Treviño 1999). When formal procedures state one thing, and informal culture recommends the opposite, organizations display a lack of alignment (Treviño et al. 1999; Weaver & Treviño 1999) and send mixed signals to employees regarding desired ethical behavior (Constantinescu and Kaptein 2015a).

Organizational values are thus applied inconsistently, creating disconnect between espoused and enacted values (Silverman 2000).

To respond to these challenges in practice, HCOs should have integrated ethics departments that manage both the implementation of ethical rules that ensure organizational compliance and the development of an ethical culture that promotes organizational integrity. Such a department would integrate and balance stakeholder concerns and expectations within the organizational environment, including rules and values. In doing this, attention should be paid to the fact that the organizational formal structure is designed independently of the individuals fulfilling specific job functions, while the informal culture is dependent on the personal characteristics of individuals (Boyle et al. 2000). From a normative stance, this means that the integrated ethics department may be approached based on different theories. On the one hand, a rational systems approach is appropriate for matters pertaining to the organizational structure, stressing that written policies and procedures are needed to support employees' ethical behavior. On the other hand, a natural or open systems approach is more appropriate for issues related to the organizational informal culture, highlighting that ethical conduct also needs moral guidance based on values. Such a systems perspective allows us to picture organizational ethics as managing the network of structured and informal relationships among multiple parties, thus "acknowledging the interdependence of all the stakeholders in the organization, internal and external: the clinicians and administration, the board, the patients and the community" (Rorty et al. 2004, 86).

A normative evaluation of the way ethics is managed at the organizational level thus becomes synonymous to the way "the actual corporate context stimulates and facilitates employees to realize the justified and fundamental expectations of stakeholders and to balance conflicting expectations in a responsible way" (Kaptein 1998, 58). This means taking into account not only the way the organization fosters ethical relations between its own members, but also the broader societal context of managing ethical issues. For HCOs, this translates into the way optimal alignment is achieved between formal medical protocols and general norms that govern clinical practice, nonmedical operations, and informal practices. Furthermore, the quality of this alignment determines the HCO's success in finding the optimal balance between expectations from multiple stakeholders.

8.5 Conclusion and suggestions for future research

Ongoing changes in the way HCOs operate have raised new issues about contextual pressures that individual members face, such as time constraints, cost efficiency, and staffing reductions (Austin 2007; Carney 2011; Wesorick 2002). This often leads to the moral distress of physicians and staff in cases when individuals can hardly live up to moral standards because of contextual issues (Austin 2007; Jameton 1993).

Professionals working in HCOs report difficulties dealing with cases that present a lack of consistency between patients' needs and organizational contextual requirements (Hart 2005). Therefore, more emphasis is needed on the HCO itself.

This article has suggested a possible means to dissolve the apparent antithesis between compliance and integrity approaches to ethical issues in HCOs operating especially in European and North American countries. I have argued that an integrative view, where compliance and integrity are on the same continuum of moral complexity, offers more explanatory power and is better able to advance research in HCO ethics. Moreover, I have emphasized that this complementary relationship mirrors two organizational dimensions: the informal culture and formal structure. HCOs must work to better align these two dimensions if they want to promote integrity in handling both internal and external stakeholder legitimate requests.

Currently, the main challenge for HCOs is to "maintain and if possible to improve the quality of care in the face of cost containment" (Rorty 2000, 59) because of the separation between cost management and clinical management, often translated into the rivalry between cost and quality. To resist the temptation to sacrifice the quality of care in favor of cost reduction, HCOs should take all efforts to make compliance and integrity complementary, thus aligning organizational culture and structure. Indeed, to be able to optimally balance legitimate stakeholder concerns related to cost and quality, organizational ethics "needs to embrace both substance and form, substance driving form, meaningfully informed by context" (Bean 2011, 325).

The need for substantive organizational ethics programs is even more urgent today given that HCOs cannot rely only on imposed legal regulations to integrate moral concerns of its multiple stakeholders. Morality goes beyond legal requirements (Constantinescu and Kaptein 2015a; Jeurissen 2004; Paine 1994) because the law addresses the minimum moral standards to be obeyed, without necessarily driving excellence or virtue. The organizational context needs to receive the importance it deserves in health care, given that formal structure and informal culture influence employee morality (Kaptein and Wempe 2002). Given that an ethical climate improves both the quality of medical care and the overall organizational performance (Suhonen 2011), research and implementation of organizational ethics in health care become utterly relevant.

Future research on HCO ethics should take a closer look at and collaborate with business ethics research. The field of ethics management could provide additional normative insights that would advance research on organizational ethics. To that end, the first suggestion for future research is to explore the full implications of the integrated model developed by Jeurissen (2004) and discuss how the moral complexity continuum may be applied to realistic health-care situations. Second, future research must analyze the implications of the lack of alignment between culture and structure on the medical act performed in HCOs and synthesize possible remedies. For instance, one such remedy proposed in business ethics research that could be applied to HCO ethics concerns the degree to which HCOs integrate ethical virtues, such as those proposed by the Corporate Ethical Virtues Model advanced by Kaptein (1998, 2008). Third, it would be critical to discuss moral criteria to ascribe blame and praise at the individual and also group levels in HCOs. In this vein, one possible line of research would be to explore the interaction between various levels of moral responsibility in HCOs and see under what circumstances this interaction may lead to a mutually enhancing responsibility (Constantinescu and Kaptein 2015a). Equally important, future research would need to explore normative and empirical methods of evaluating the way ethics is managed at the organizational level in health care. Finally, more attention should be paid to the philosophical underpinnings of organizational ethics. Developments of virtue-based ethics approaches to business ethics (Kaptein 1998, 2008; Moore 2012, 2015; Solomon 1992, 2004) could potentially offer a relevant anchor of understanding moral issues at the organizational level in health-care settings.

References

Austin, Wendy. 2007. The Ethics of Everyday Practice Healthcare Environments as Moral Communities. In *Advances in Nursing Science*, 30(1): 81–88.

Barina, Rachelle. 2014. Ethics Outside of Inpatient Care: The Need for Alliances between Clinical and Organizational Ethics. In *HEC Forum*, 26(4): 309–323.

Bean, Sally. 2011. Navigating the Murky Intersection between Clinical and Organizational Ethics: A Hybrid Case Taxonomy. In *Bioethics*, 25(6): 320-325.

Boyle, Philip J., Edwin R. DuBose, Stephen J. Ellingson, David E. Guinn, and David B. McCurdy. 2000. *Organizational Ethics in Health Care: Principles, Cases, and Practical Solutions*. Jossey-Bass: San Francisco.

Carney, Marie. 2011. Influence of organizational culture on quality healthcare delivery. In *International Journal of Health Care Quality Assurance*, 24 (7): 523 – 539.

Chen, Donna T., Patricia H. Werhane, and Ann E. Mills. 2007. Role of organization ethics in critical care medicine. In *Critical care medicine*, 35(2) Suppl: S11-7.

Constantinescu, Mihaela, and Muel Kaptein. 2015a. Mutually Enhancing Responsibility: A Theoretical Exploration of the Interaction Mechanisms between Individual and Corporate Moral Responsibility, in *Journal of Business Ethics*, 129 (2), 325-339.

Constantinescu, Mihaela, and Muel Kaptein. 2015b. *CSR Standards and Corporate Ethical Virtues: A Normative Inquiry into the Way Corporations Integrate Stakeholder Expectations*, in Samuel O. Idowu et. al. (Eds.), *Corporate Social Responsibility and Governance. Theory and Practice*, Switzerland: Springer, 159-180.

Emanuel, Ezekiel J. 1995. On the Need for Institutional Structures Instead of Principles for Individual Cases. In *The Journal of Clinical Ethics*, 6(4): 335-338.

Førde, Reidun, and Thor Willy Ruud Hansen. 2014. Do Organizational and Clinical Ethics in a Hospital Setting Need Different Venues? In *HEC Forum*, 26: 147–158.

Fox, Ellen, Melissa M. Bottrell, Kenneth A. Berkowitz, Barbara L. Chanko, Mary Beth Foglia, and Robert A. Pearlman. 2012. Integrated ethics: An innovative program to improve ethics quality in health care. In *The Innovation Journal: The public Sector Innovation Journal*, 15(2):1–29.

Gallagher, J.A., and J. Goodstein. 2002. Fulfilling institutional responsibilities in health care: Organizational ethics and the role of mission discernment. In *Business Ethics Quarterly*, 12 (4): 433-450.

Gibson, J.A. 2012. Organizational ethics: No longer the elephant in the room. In *Healthcare Management Forum*, 25: 37–39.

Gibson J.L., Connolly E., Sibbald R., et al. 2008. *Organizational ethics*. In: Singer P.A., Viens A, (Eds.), *The Cambridge Textbook of Bioethics*. Cambridge: Cambridge University Press: 243–250.

Hall, RT. 2000. *An Introduction to Health Care Organizational Ethics*. Oxford: Oxford University Press.

Hart, S.E. 2005. Hospital ethical climates and registered nurses' turnover intentions. In *Journal of Nursing Scholarship*, 37(2): 173–7.

Jeurissen, R. 2004. *Moral complexity in organizations*. In: M. Korthals, R. Bogers (Eds.), *Ethics for life scientists*. Springer: Dordrecht, 11-20.

Jameton, A. 1993. Dilemmas of moral distress: moral responsibility and nursing practice. In *AWHONN's Clinical Issues in Perinatal Women's Health Nursing*, 4: 542–551.

Joint Commission on Accreditation on Healthcare Organizations. 1996. *Patient rights and organizational ethics: Standards for organizational ethics*. In: *Comprehensive manual for hospitals*. Oakbrook Terrace, IL.

Kaptein, Muel. 2011. Understanding Unethical Behavior by Unravelling Ethical Culture. In *Human Relations*, 64(6), 843-869.

Kaptein, Muel. 2008. Developing and Testing a Measure for the Ethical Culture of Organizations: The Corporate Ethical Virtues Model. In *Journal of Organizational Behaviour*, 29(7), 923-947.

Kaptein, Muel, and Wempe, J. 2002. *The Balanced Company: A Theory of Corporate Integrity*. New York, Oxford University Press.

Kaptein, Muel. 1998. *Ethics Management: Auditing and developing the ethical content of organizations*. Dordrecht: Kluwer Academic Publishers.

Khushf, G. 1998. The Scope of Organizational Ethics. In *HEC Forum*, 10(2): 127-135.

Magill, G., Prybil, L. 2004. Stewardship and Integrity in Health Care: A Role for Organizational Ethics. In *Journal of Business Ethics*, 50: 225–238.

Mills, A.E., Spencer, E.M. 2001. Organization Ethics or Compliance: Which will Articulate Values for the United States' Healthcare System? In *HEC Forum*, 13(4): 329-43.

Paine, L. S. 1994. Managing for Organizational Integrity. *Harvard Business Review*, 72(2), 106-117.

Phillips, R. A., Margolis, J. D. 1999. Toward an Ethics of Organizations. In *Business Ethics Quarterly* 9(4): 619-638.

Potter, R.L. 1996. From clinical ethics to organizational ethics: The second stage of the evolution of bioethics. In *Bioethics Forum*, 12: 3-12.

Reiser S. 1994. The ethical life of health care organizations. In *Hastings Cent Rep.*, 24: 28–36.

Rorty, M.V. 2000. Ethics and Economics in Healthcare: The Role of Organization Ethics. In *HEC Forum*, 12(1): 57-68.

Rorty, M.V., Werhane, P.H., Mills, A.M. 2004. The Rashomon effect: organization ethics in health care. In *HEC Forum, 16(2): 75-94.*

Silverman, H. J. 2000. Organizational Ethics in Healthcare Organizations: Proactively Managing the Ethical Climate to Ensure Organizational Integrity. In *HEC Forum* 12(3): 202-215.

Solomon, R. C. 2004. Aristotle, Ethics and Business Organizations. In *Organization Studies*, 25(6): 1021-1043.

Solomon, R. C. 1992. Corporate Roles, Personal Virtues: An Aristotelean Approach to Business Ethics. In *Business Ethics Quarterly*, 2(3): 317-339.

Spencer E.M., Mills A.E., Rorty, M.V., Werhane, P.H. (Eds.). 2000. *Organization ethics in health care*. New York: Oxford University Press.

Suhonen, R., Stolt, M., Virtanen, H., Leino-Kilpi, H. 2011. Organizational ethics: A literature review. In *Nursing Ethics*, 18(3): 285–303.

Treviño, L. K., Weaver G.R., Gibson D.G., et al. 1999. Managing ethics and legal compliance: What works and what hurts? In *California Management Review,* 41(2): 131–151.

Treviño L. K., Youngblood, S.A. 1990. Bad apples in bad barrels: A causal analysis of ethical decision-making behavior. In *Journal of Applied Psychology*, 75(4): 378-385.

Wall, S. 2007. Organizational Ethics, Change, and Stakeholder Involvement: A Survey of Physicians. In *HEC Forum*, 19 (3): 227–243.

Wesorick, B. 2002. 21st century leadership challenge: creating and sustaining healthy, healing work cultures and integrated service at the point of care. In *Nursing Administration Quarterly,* 26: 18-32.

Weaver G., Treviño L. K. 1999. Compliance and values oriented ethics programs: Influences on employees' attitudes and behaviors. In *Business Ethics Quarterly,* 9: 325–345.

Werhane, P.H. 2000. Business Ethics, Stakeholder Theory, and the Ethics of Healthcare Organizations. In *Cambridge Quarterly of Healthcare Ethics*, 9: 169–181.

Eva De Clercq

9 Disability @ the Movies: Toward a Disability-Conscious Bioethics

What continues to concern me most is what will and what will not constitute an intelligible life, what will and will not be considered to be "real".
(J. Butler, *Gender Trouble* 1990)

9.1 Introduction

The relationship between disability, cinema, and bioethics is a love–hate affair. Bioethics shares with the disability rights movement its commitment to patient autonomy over and against medical paternalism (Asch 2001, 297–9). Nevertheless, mainstream Anglo-American bioethics has shown little or no interest for the topic of disability and the impact of impairment on everyday life (Kuczewski 2001). When attention has been given, discussions have been mostly limited to quality-of-life analyses or decisions at the beginning (e.g., prenatal screening and abortion), as well as at the end of life (e.g., requests for assistance in dying) (Kuczewski 2001, 36). The reason for this is that bioethicists have focused mainly on fashionable topics or cutting-edge technologies, such as gene editing, moral enhancement, and assistive robots (Kuczewski 2001, 36). The inattention and indifference for everyday problems of people with disabilities explains the continuing tensions between bioethics and advocates of the disability rights movement (Amundson and Tresky 2008; Ouellette 2011). Unlike Anglo-American bioethics, European bioethics is much less procedural and more teleological (Schotsmans 2015). This means that it does not focus only on principles such as autonomy and nonmaleficence but is also concerned with people's dignity, integrity, and vulnerability. Unfortunately, for a long time, the presence of European bioethics on the intercontinental scene has remained rather limited, with the result that the (individualistic) autonomy-oriented approach has largely prevailed (Schotsmans 2015).

Unlike mainstream bioethics, film narratives have rarely ever ignored disability. In fact, compared to race and gender, disabled bodies are almost "obsessively" present on the screen (Chivers and Markotic 2010, 1; Mitchel and Snyder 2001, 51–2). Still, the striking fact is that despite this prevalence, the presence of disabled characters is often overlooked by the audience in the sense that viewers seldom recognize disability as a feature of the film (Longmore 2003). The reason for this is that disability is used as an (invisible) narrative "prosthesis": stories rely on disability for its melodramatic power to evoke emotions of fear and compassion but rarely ever focus on it as an experience of social, cultural, and political dimensions (Mitchell and Snyder 2001, 48).

In other words, both cinema and bioethics make disabled bodies invisible. Or yet, they exclude them from the boundaries of the real: of those whose lives can be counted as culturally viable. The American philosopher and gender theorist Judith Butler calls this process a regulatory mechanism of de-realization (Butler 2004). This can occur in two ways: through omission (exclusion) or through the mode of visual representation itself (Butler 2004, 147). Although, Butler's work does not explicitly deal with disability (Samuels 2002, 59–61)[1], her theoretical framework constitutes a valuable key to better understand the problematic relationship between disability, bioethics, and cinema. Disabled bodies become unreal in bioethics due to exclusion and, in cinema, due to the way in which they are represented.

In this paper, I argue that it is possible to come to a disability-conscious bioethics (Ouellette, 2011) through cinema. Films can play an important role in bioethics curriculum as they can change the way in which future health professionals and scholars within bioethics look at people with disabilities. This is especially the case for the educational system within the European tradition, wherein bioethics is not just a method for ethical decision-making but is also guided by an anthropological approach that uses ethical principles to protect the development of the human character (Adachi and ten Have 2015; Schotsmans 2015).

I want to explore the possibility of such a disability-conscious bioethics by revisiting Martha Nussbaum's notion of moral imagination (1990) and Butler's concept of re-signification (1997). I start with illustrating the individual and social models of disability through a brief discussion of a film fragment with Judith Butler and disability rights activist Sunaura Taylor. In the next section, I explore Nussbaum's work on the role of literature in the cultivation of our empathy. I then transpose this framework to cinema. I examine the stories and cinematographic language that have encoded disability throughout the history of film. In the last section, I focus on the portrayal of disability in more recent films and I argue, following Butler, that images can be re-signified or interpreted in unexpected ways and, thus, give rise to alternative meanings of disability.

1 Butler is a professor in the Department of Comparative Literature and the Program of Critical Theory at the University of California, Berkeley, CA, USA. She has made significant contributions to gender-, sexuality-, and feminism-related issues. Still, in her most recent work, she focuses also on human rights, antiwar politics, and mourning.

9.2 What *can* a body *do*?

Examined Life (Astra Taylor 2008) is a documentary film featuring eight living philosophers, among whom is Judith Butler. We see her "going for a walk" together with Sunaura Taylor in the streets of San Francisco, one of the most disabled-friendly cities in the world. Taylor was born with a neuromuscular disorder and sits in a wheelchair. Butler and Taylor start their conversation by considering the question: what *can* a body *do*? This is a quite unusual question in the history of philosophy (Abrams 2011, 74). In fact, most philosophers have occupied themselves with *what* the body *is* (Abrams 2011, 74). Both feminists and disability activists have viewed the latter question with suspicion, as the discrimination of women and people with disabilities has often been justified by referring to their "weak" and "impaired" bodies. Asking what the body can *do* draws attention away from the essence of the body and shifts the focus to capabilities and possibilities (Abrams 2011, 74–5). Still, in the video fragment, two different meanings are given to this question: "what is a body *able* to do?" and "what is a body *permitted* to do?" (Abrams 2011, 75).

The first formulation brings Taylor to the difference between the medical and the social models of disability. The medical model views disability as a lack of ability to perform *normal* human activities due to physical or mental impairments. Disability is defined as a *personal* problem that needs to be managed individually by providing people special treatment and services. Rehabilitation medicine is meant to alleviate discomfort and restore functioning (Banja 2015, 562–3) to allow a person to lead a satisfying life. Wheelchairs or prosthetic limbs, for instance, correct persons' impaired bodies and enable them to function "normally". Traditionally, bioethicists and policymakers have adhered to the medical model of disability (Asch 2001).

The social model presents a radically different view on disability. The inability to participate in society is no longer considered to be a personal issue but collocated in societal and environmental barriers that inhibit people with impairments from leading a "normal" life. Examples of social disabling restrictions are limited access to health care, education, employment, public spaces, independent living, and transport, as well as the stereotypical images conveyed by cultural products, such as films, TV, social media, and so on. This inaccessibility reinforces social unacceptability. Taylor gives us a concrete idea of this social discrimination: she remembers that when she was little and "walked" without her wheelchair, people compared her to a monkey because her body was perceived as transgressing the norm of ableism (Abrams 2011, 79). This example clearly shows that there are important social constraints to people's ability to move and that the question "what *can* a body do?" is closely connected to issues of normalization.

This brings us to the second formulation: what is a body *permitted* to do? (Abrams 2011, 79–80). This question has to do with societal norms and expectations regarding embodiment, which limit what can be considered to be real and acceptable. Taylor challenges these norms, for instance, by using her mouth for activities for which other people use their arms and hands, such as cooking, cleaning, painting, and even picking up a coffee cup, and this causes discomfort in an "ableist" society as ours (Abrams 2011, 89). On the other hand, asking for assistance with basic needs is a provocative act as it challenges the assumption of autonomy, independence, and self-reliance. In their conversation, Butler and Taylor interrogate these dominant norms of "able-ism" and emphasize the importance of recognizing vulnerability and interdependence as a common human condition of both "able" and "disabled" persons (Abrams 2011, 85–9). They remind us that disability is a fundamental part of human life and that, sooner or later, most of us will experience some kind of disability in our lives.

9.3 Moral imagination @ the movies

In our contemporary society, images are everywhere. The power of visual culture on public opinion is much greater than that of words. This may explain why Italy's national organization for people with Down syndrome decided to launch a video on World Down Syndrome Day to raise awareness for this condition (Coordown 2016). The film *How do you see me* takes the form of an internal monologue of a young woman who sees herself as an ordinary person who goes for a run, works in a restaurant, hangs out with her family and friends, laughs, and cries. The video features the American actress Olivia Wilde, but the voice is of Anna Rose Rubright, a young woman with Down syndrome, who only appears at the very end of the movie to ask: "This is how *I* see myself, how do *you* see me?". The project aimed to overcome the stereotypical view that people with this condition cannot lead fulfilling lives. Although the film was reviewed positively by parents and various media, there were also critical voices (Perry 2016). According to some disability activists, the fact that Anne Rose is visually erased reinforces the idea that disabled bodies should not be seen ("unreal") and that disabled persons (should) see themselves through the (ideal) eyes of able-ism. Many people with disabilities have been confronted with statements such as, "I do not think of you as disabled, you're my friend!" or "I don't see you as disabled, I just see you as a person!" (Garland-Thomson 2016; Hitselberger 2016). Despite the best intentions, these words risk undermining the value of disabled lives. Karin Hitselberger (2016) puts it as follows:

> How you see me matters. See me as me. See me as a daughter, sister, friend, writer, and student. See me as smart, strong, outgoing, and capable. See me as all these things, but see me as disabled, too. See my chair, and acknowledge that it changes the way I experience the world. See my disability, and understand that is an integral part of who I am. See me, and realize that I don't have to erase my disability in order for any other part of me to shine through.

The main difference between Sunaura Taylor and Anne Rose is that where the former emerges as a woman, a thinker, and an activist with physical impairments (Abrams 2011, 87), the latter emerges as a person *despite* the fact that she is disabled. Although Taylor is not defined by her physical disabilities, as viewers, we are nevertheless aware of the fact that she has limited movement in arms and legs (Abrams 2011, 87–9). At the same time, however, we do not see her as a disabled person trying to become abled. She has her own, distinct abilities (Abrams 2011, 87–99). We can imagine how it is to live with a neuromuscular disorder: it not only affects how she moves but also how she interacts with people around her.

This leads us to the need for moral imagination. In *From Disgust to Humanity: Sexual Orientation and Constitutional Law* (2010, xvii), Martha Nussbaum writes:

> It is possible to view another human being as a slimy slug or a piece of revolting trash only if one has never made a serious good-faith attempt to see the world through that person's eyes or to experience that person's feelings. Disgust imputes to the other a subhuman nature. How, by contrast, do we ever become able to see one another as human? Only through the exercise of imagination.

Nussbaum argues that moral imagination is encouraged by the activity of novel reading (1990, 166). Literature is nourishing because it expands our empathy and knowledge: it gives us access to the thoughts, feelings, and dreams of those different from ourselves; it enables us to live multiple lives and to take on their viewpoints. Afterward, it becomes more difficult to find these "others" disgusting or abhorrent. This is why reading a novel is never a trivial activity; on the contrary, it is subversive (Nussbaum 1991, 879): it urges the reader to question dogmatic conventions, to challenge social prejudices, and to move beyond preconceived generalizations. For the same reason, Nussbaum believes that literature has an important role to play in public life: it "will steer judges in their judging, legislators in their legislating and policymakers in measuring the quality of life of people both near and far" (Nussbaum 1991, 879).

Nussbaum's invitation to enlarge our moral imagination by reading novels should not be interpreted as a rejection of traditional ethical theories; on the contrary, literature and moral theory are complementary. Still, literature is not the only vehicle to stimulate moral imagination. Similar to novels, films allow people to approach moral questions in various directions as they offer us an endless amount of stories and lives to experience. The interest in cinema as a source to strengthen our moral capacities is evidenced by the growing number of authors publishing on this topic. Further, various movie festivals and university courses incorporate feature films in the student curriculum (Penn, Stanford, and Yale Universities) in order to encourage and enhance students' thinking about important ethical issues (DiBartolo and Seldomridge 2009).

9.4 Disability stereotypes on the silver screen

Cinema and disability constitute a strange combination for many of us. Still, as stated above, disability has never really been absent from the silver screen. The problem is not that people with disabilities have been underrepresented, but rather that the film industry has conveyed a pronounced stereotypical image of these people by showing little or no interest in disability as a condition of life. They have been – to use a term of Judith Butler – de-realized. In *The Cinema of Isolation* (1994), the American scholar Martin Norden puts it as follows:

> The movie industry has perpetuated or initiated a number of stereotypes over the years as part of the general practice of isolation—stereotypes so durable and pervasive that they have become mainstream society's perception of disabled people and have obscured if not outright supplanted disabled people's perceptions of themselves (Norden 1994, 3).

Until the 1990s, people with disabilities were rarely ever featured as protagonists but were set apart from the other characters in the film to highlight a physical and symbolic separation between them and the rest of the "abled" society (Norden 1994, 3–4). This process of isolation is not only enacted by the plot of the film but also by the so-called cinematographic language (framing, lighting, editing, and sound) (Norden 1994; Ellis 2008, 35).

In his work, Norden traces the history of the way in which physical disability has been represented in American films. He distinguishes three historical periods: (1) from the birth of cinema to the end of the 1930s; (2) from World War II (WWII) until the 1970s; and (3) from the 1970s until the mid-90s (when his work was published). In the first period, people with disabilities were depicted as monsters, criminals, or comic characters whose bodies were used to arouse pity, compassion, and fear in the audience (*The Hunchback of Notre Dame*, Wallace Worsely, 1923; *Freaks*, Browning, 1932). Many movies conveyed the message that these people should be excluded from society because their deformed bodies were considered to be a reflection of the deformation of their soul. For those with a permanent disability, there was no other way out but death. This might also explain the interest in storylines that revolved around characters with curable disabilities (*City Lights*, Charlie Chaplin 1931). After WWII, the return of many disabled veterans, young men who had put their lives at risk to protect the nation, urged a different, more sensitive representation of disability. As a result, films from the mid-1940s onward portrayed people with disabilities no longer as criminals or freaks but as people who were engaged in a personal battle (*The Men*, Fred Zinneman 1950), much like the medical model of disability. They insisted on the need to fix disabled people, instead of making society more inclusive. The implied message was that reintegration and success in society are only possible when based on the tenacity to overcome adversity and that self-pity can only lead to social exclusion (Norden 2014).

Films belonging to the third period were not so much concerned with the theme of rehabilitation but focused more on social struggles. This shift in representation should be understood in the context of the Rehabilitation Act of 1973, which ensured equal opportunities for people with disabilities and gave rise to the so-called social model of disability. Still, despite this progress, many films of the 1990s still took a rather paternalistic and normalizing approach. In *Rain Man* (Barry Levinson 1988), for instance, Raymond's autism is somehow compensated by the supernatural gift of his phenomenal memory. Likewise, Tom Hanks' mental retardation in *Forrest Gump* (Robert Zemeckis 1994) turns him into a kind of spiritual guru. Still, the film portrays Forrest as an inherent asexual being, continuing the trope that people with disabilities are not – and should not be – sexually active.

Throughout his book, Norden thus shows us cinema's long history of stereotyping disability and urges the audience to reflect on these stereotypes, which are at once the *result* and the *cause* of society's negative attitudes toward disabled people. These preconceived notions enact a kind of normative "violence": although disability is represented, the way in which it is framed ("put on screen") inhibits us from imagining it as a meaningful and viable ("real") way of life. Or yet, persistent sociocultural norms, such as autonomy and invulnerability, have enabled us to turn disabled bodies into anomalous, deviant bodies that do not matter. Given this violence of de-realization in films, we should reflect on the interaction between film and society and ask ourselves what role – if any – films can have today in challenging these stereotypes.

9.5 Disability in contemporary cinema: from isolation to inclusion?

More than 2 decades have passed since the publication of Norden's famous book. Since then the field of disability studies has grown consistently (Jarman and Kafer 2014). This may explain why the American president Donald Trump sparked a worldwide outrage during one of his election campaigns, when he mocked a disabled journalist by making spastic movements with his arms. This awkwardness in laughing at disabled people is not just a matter of political correctness but shows a change of mind-set regarding how people perceive disability. Still, this does not mean that societal prejudices have disappeared or that disability is no longer used as a trope to make fun of people in the media. Hence, it is important "to trace out the longstanding tradition of representational strategies" (Snyder and Mitchell 2010, 195) that continue to denigrate people with disabilities. Although many scholars have criticized the negative *narrative* portrayal of disability in movies, much less attention has been given to cinematographic *language* and the relationship between the audience and the actor's *body* (Snyder and Mitchell 2010, 181). Films, in fact, heavily rely on the representational power of the body. This is especially the case for those genres, such as comedies, melodrama, and horror movies, for which success depends on the capacity to generate sensations in the bodies of their viewers (Snyder and Mitchell 2010, 183; Williams 1991, 702). In these so-called "body genres", the disabled body plays the role of (1) producer of fear and trauma (horror), (2) the victim of bodily disintegration (illness melodrama), and (3) the metaphor of loss of bodily control (comedy) (Snyder and Mitchell 2010, 186). This need to communicate intense emotions to the audience might explain the long-term, almost obsessive, interest of the film industry in disability, especially in the period of the origins when cinema had few other means at its disposal to evoke strong reactions.

This raises the question whether films of recent years have managed to overcome this visual and narrative rhetoric and are "finally moving with times" (Cox 2012). Let us start with *The Sessions* (Lewin 2012). The movie tells the story of journalist and poet Mark O'Brien, who was paralyzed by polio as a child and is since then confined to an iron lung. At the age of 38, he decides to lose his virginity and hires a sex therapist to initiate him into the world of sexuality. The film has the merit of tackling the taboo of sex and disability but the comedy genre risks masking the complexities that disabled people face in the sphere of intimacy. *Rust and Bone* (Audiard 2012) tells the love story of Ali, a single father with no money, and Stephanie, a young woman who loses both legs after a tragic accident while training orcas in a tourist park. In the sex scenes, Stephanie's impaired body is shown quite explicitly (although they are computer-generated images), but the movie's true focus is more on Ali's emotional disability. Despite Stephanie's gradual self-empowerment, the film has been criticized for minimizing the difficulties of being a double-amputee woman and for ignoring the fact that disabled women are often targeted as easy sexual preys (Shapiro 2013). In Michael Haneke's *Amour* (2012), George takes care of his wife Anne, who has had a stroke and gets worse every day. The movie has the merit of associating disability with old age, highlighting the fact that in an aging society such as ours, we will all become disabled one day. Although the couple's love seems to be able to face all difficulties, in the end, the only way out of Anne's permanent disabling disease is death. Likewise, in the much-debated movie *Me before You* (Sharrock 2016), the romance between Will – a paralyzed young banker – and Louisa, his caregiver, is not strong enough to withhold Will from his desire for assisted dying. For many disability activists, the message of the films seems to be that disabled people are better dead as their lives are not worth living.

Many other recent films have disability at the center of the plot, to mention just a few: *Avatar* (Cameron 2009); *Still Alice* (Glatzer 2014); *The Theory of Everything* (Marsh 2014); and *The Bélier Family* (Lartigau 2014). They all have received both positive and negative critiques from the disability community because although there – finally – seems to be a real interest in the lived reality of people with disabilities, the way in which their lives are portrayed is not always as complex and realistic. Moreover, most of the time, the disabled character is played by a nondisabled actor. In this way, "the audience can rest comfortably assured that the central character may appear to be disabled but isn't really a disabled person" (Davis 2013, 40) and can, in fact, go back to his or her "normal" life. And as many scholars have pointed out, playing a disabled character is one of the safest ways for abled actors to get an Oscar nomination for nondisabled actors (Chivers and Markotic 2010, 6), probably because of the effort that interpreting such a character requires.

Nevertheless, there are some important exceptions (Davis 2013, 40). In 2014, the Ukrainian film *The Tribe* (Slaboshpytskiy 2014) was considered to be one of the most controversial and challenging disability movies ever made. Acted entirely by deaf actors in Ukrainian sign language and without subtitles, the film forces the audience to focus on the characters' expressions and perceive communication in a new way. Still, although the language of the film is innovative, the plot – a criminal gang engaged in robbery and prostitution in a boarding school for deaf children – repeats the trope of violence and disability and does not tackle the real issues that deaf children face.

9.6 Conclusion: the power of re-signification

The relationship between disability and the film industry has changed considerably over the past 120 years. Characters with disabilities are no longer just a prosthetic vehicle of emotion (isolation) but have taken center stage in the main plot (inclusion). This may explain why, unlike in the past, disability is increasingly recognized as a dominant film feature by the audience. This does not mean that all stereotypes have disappeared: in many movies, disability is still framed as a challenge that needs to be overcome, whether through the plot (the disabled character has a special gift) or the cinematographic language itself (use of nondisabled actors). Still, overall, both the narrative and the language have become more complex in the sense that they move beyond the monotone characterization of disabled people as villains, victims, or heroes. Of course, one could always ask for a "truer" image – although one could question to what extent fictional stories should adhere to the truth – but the fact that the lives of people with disabilities are put onto screen should not be underestimated. Following Butler's logic of re-signification (1997), I believe that viewers can interpret movies in many unexpected ways that are beyond the control of the filmmakers. This is testified by the various contrasting viewpoints about the movies analyzed in the previous section. These discussions are fruitful in challenging stereotypical frames of representation and therefore movies can – despite all their limitations – encourage our moral imagination of what it is like to live with a disability. Hence, they can play an important role in bioethics education, which is increasingly moving from problem-solving toward influencing students' attitudes, behaviors, and characters (Adachi and ten Have 2015, 7). Changing the way in which future health professionals and scholars within bioethics look at people with disabilities is an important step to arrive at a more disability-conscious bioethics.

References

Abrams, Kathryn. 2011. "Performing Interdependence: Judith Butler and Sunaura Taylor in the Examined Life," *Columbia Journal of Gender and Law* 21: 72-89.
Adachi, Toshitaka, and Henk ten Have. 2015. *Bioethics education in a global perspective*. Dordrecht: Springer.
Banja, John D. 2015. "The Disability Movement's Critique of Rehabilitation's Medical Model: A Rebuttal." *American Medical Association Journal of Ethics* 17(6): 562-567.
Amundson, Ron, and Shari Tresky. 2008. "Bioethics and Disability Rights: Conflicting Values and Perspectives." *Bioethical Inquiry* 5: 111-123.
Asch, Adrienne. 2001. "Disability, Bioethics, and Human Rights." In *Handbook of Disability Studies*, ed. Gary L. Albrecht, Katherine Seelman and Michael Bury. Thousand Oaks, California: Sage Publications, 297-326.
Audiard, Jacques. 2012. *Rust and Bone*.
Browning, Tod. 1932. *Freaks*.
Butler, Judith. 1990. *Gender trouble: feminism and the subversion of identity*. New York: Routledge.
Butler, Judith. 1997. *Excitable speech: a politics of the performative*. New York: Routledge.
Butler, Judith. 2004. *Precarious life: the powers of mourning and violence*. London: Verso.
Cameron, James. 2009. *Avatar*.
Charlie, Chaplin. 1931. *City Lights*.
Chivers, Sally, and Nicole Markotić. 2010. *The problem body: projecting disability on film*. Columbus: Ohio State University Press.
"Coordown: How do you see me? March 21 World Down Syndrome Day", last modified March 14, 2016, http://www.coordown.it/www/index.php.
Cox, David. November 12, 2012 (12.25). "Disability in film: is cinema finally moving with the times?" *The Guardian Film Blog*, https://www.theguardian.com/film/filmblog/2012/nov/12/disability-in-film-untouchable.
Davis, Leonard J. 2013. *The end of normal identity in a biocultural era*. Ann Arbor, Mich: The University of Michigan Press.
DiBartolo, Mary, and Lisa Seldomridge. 2009. "Cinemeducation: Teaching end-of-life issues using feature films." *Journal of Gerontological Nursing* 35(8): 30-36.
Ellis, Katie. 2008. *Disabling Diversity. The Social Construction of Disability in 1990s Australian National Cinema*. Saarbrücken: VDM Verlag Dr. Müller.
Garland-Thomsen, Rosemarie. August 19, 2016. "Becoming Disabled. Roughly one in five Americans lives with a disability. So where is our pride movement?" *The New York Times Opinion*, https://www.nytimes.com/2016/08/21/opinion/sunday/becoming-disabled.html?_r=3.
Glatzer, Richard. 2014. *Still Alice*.
Haneke, Michael. 2012. *Amour*.
Hetselberger, Karin. March 24, 2016 (08:26 am). "Why it's Important My Wheelchair is Part of How You See Me." *The Huffington Post*, http://www.huffingtonpost.com/karin-hitselberger/why-my-wheelchair-is-important-in-how-you-see-me_b_9523778.html.
Jarman, Michelle, and Alison Kafer. 2014. "Guest Editors' Introduction: Growing Disability Studies: Politics of Access, Politics of Collaboration." The Disability Studies Quarterly. The first journal in the field of disability studies 34(2), http://dsq-sds.org/article/view/4286/3585.
Kuczewski, Mark. 2001. "Disability: An Agenda for Bioethics." *American Journal for Bioethics* 1(3): 36-44.
Lartigau, Eric. 2014. *The Bélier Family*.
Levinson, Barry. 1988. *Rain Man*.
Lewin, Ben. 2012. *The Sessions*.

Longmore, Paul K. 2003. *Why I burned my book and other essays on disability*. Philadelphia: Temple University Press.

Marsh, James. 2014. *The Theory of Everything*.

Mitchell, David T., and Sharon L. Snyder. 2001. *Narrative prosthesis: disability and the dependencies of discourse*. Ann Arbor: University of Michigan Press.

Norden, Martin F. 1994. *The cinema of isolation: a history of physical disability in the movies*. New Brunswick, N.J.: Rutgers University Press.

Nussbaum, Martha C. 1990. *Love's knowledge: essays on philosophy and literature*. New York: Oxford University Press.

Nussbaum, Martha C. 1991. "The Literary Imagination in Public Life." *New Literary History* 22(4): 877-910.

Nussbaum, Martha Craven. 2010. *From disgust to humanity: sexual orientation and constitutional law*. Oxford: Oxford University Press.

Ouellette, Alicia. 2011. *Bioethics and disability: toward a disability-conscious bioethics*. Cambridge: Cambridge University Press.

Perry, David. March 21, 2016. "The Problem with the New Down Syndrome PSA Starring Olivia Wilde." *The Establishment*, https://theestablishment.co/the-problem-with-the-new-down-syndrome-psa-starring-olivia-wilde-bafe20e163e9#.w5of1hvg1.

Samuels, Ellen. 2002. "Critical Divides: Judith Butler's Body Theory and the Question of Disability." *NWSA Journal* 14(3): 58-76.

Schotsmans, Paul. 2015. "Bioethics past, present and future: a personal and narrative perspective from the European continent." In *The voices and rooms of European Bioethics*, edited by Richard Huxtable and Ruud ter Meulen, 17-30. London: Routledge.

Shapiro, Lawrence. 2013. "Cinematic Impressions of the Female Amputee." *Disability Studies Quarterly* 33, 2. Accessed April 14, 2016. http://dsq-sds.org/article/view/3595/3241.

Shapshay, Sandra. 2009. *Bioethics at the movies*. Baltimore: Johns Hopkins University Press.

Sharrock, Thea. 2016. *Me before You*.

Slaboshpytskiy, Myroslav. 2014. *The Tribe*.

Snyder, Sharon, and David Mitchell. 2010. "Body Genres: An Anatomy of Disability in Film." In *The problem body: projecting disability on film*, ed. Sally Chivers and Nicole Markotić. Ohio State University Press, 179-206.

Taylor, Astra. 2008. *Examined Life*.

Williams, Laura. 1991. "Film Bodies: Gender, Genre and Excess." *Film Quarterly* 44(4): 2-13.

Worsely, Wallace. 1923. *The Hunchback of Notre Dame*.

Zemeckis, Robert. 1994. *Forrest Gump*.

Zinneman, Fred. 1950. *The Men*.

Constantin Vică

10 The Info-Computational Turn in Bioethics

10.1 All watched over by machines of loving grace?

In 1967, one of the most eccentric American writers, Richard Brautigan, wrote a visionary poem about the future of our species in a cybernetic world:

> I like to think (and/the sooner the better!)/of a cybernetic meadow/where mammals and computers/live together in mutually/programming harmony like pure water/touching clear sky. [...] I like to think/(it has to be!)/of a cybernetic ecology/where we are free of our labors/and joined back to nature,/returned to our mammal/brothers and sisters,/and all watched over/by machines of loving grace.

This visionary poem was revived in a British Broadcasting Corporation (BBC) documentary by Adam Curtis because it encompasses our deepest expectations, hopes, and fears about living in a digital realm, a way of life that is becoming increasingly dystopian.

It took a long time for us to accept that we are evolved natural automata, i.e., nondeterministic but probabilistic organisms. Perhaps this idea of human beings lacks what many of us hold to be one of the most important characteristics of humanity: we are not mindless creatures but cultural beings capable of developing our own consciousness through time. We self-define as being part of a unique category that transcends the limits of the natural. Still, this intellectual reality we impose above nature is not very helpful when we encounter *other* artificial kinds.

Bioethics as a field is diverse and complex enough to escape from precise taxonomies. Although it is a perpetual activity of finding purpose, criteria, and justifications, the bioethical discourse can be understood or appraised in numerical terms. This is the reason why this blend of medical thinking and practical ethics, this combination of empirical and normative dimensions, gets much closer to computability (i.e., algorithmic treatment of information and knowledge) than practical philosophy alone. An info-computational perspective in bioethics opens the possibility that moral subjects should not be necessarily understood *as* biological entities before ascribing their metaphysical, epistemological, and moral status.

I plead at large that research and practice in medical ethics and bioethics should take into account the convergence of computing, information, and life, i.e., the convergence of artificial and natural kinds, into a single new technological lifeworld (Ihde 1990; Mitcham 2014, 25)[1]. Otherwise, the moral human agent and patient will be lost from sight and will become incomprehensible. The human being must be reaffirmed in a different manner than before: medicine and health care are becoming progressively "infocentric",[2] and this change entails a conceptual shift in what makes a human a person (and a change in the relationship of agency, duties, environment, and therapy). The last decade created the foundation for a myriad of new technologies now used in day-to-day medicine, health-care practices, and biomedical research. The development in bioinformatics made gene sequencing and editing technologies, such as clustered regularly interspaced short palindromic repeats (CRISPR)/CRISPR-associated proteins (Cas) 9, possible. Large sets of data, including genomic data, and algorithmic analysis, i.e., Big Data, open the way for new insights into both pharmaceutical research and medical practice. Automation of health care through intelligent assistive technologies is on the rise (Ienca et al. 2016; 2017). Nonetheless, the Internet is a medium for telemedicine and self-medicine: more and more people get their health information online, and this behavior is producing huge amounts of data[3]. The quantified-self movement, based on applications that monitor body functions and status, produces a variety of "small' medical data (Swan 2013), while virtual reality environments create possibilities for medical training and therapy (Opriş et al. 2012). This technological setting generates, in the words of James Moor, "policy vacuums and conceptual muddles" (Moor 2008, 34). I believe we cannot cope with them by turning to the mainstream theoretical view of agency in bioethics. Why?

The impact of these technologies on the bioethics framework of thinking will be radical and essential, rather than limited and contingent. Bioethics was structured as a field of inquiry mainly about natural beings and biological life[4] – bodies and minds, their moralities and policies – but the "fourth revolution" (Floridi 2014) introduces new entities and processes into this field of inquiry. What is different about this informational revolution resides first of all in its capacity to penetrate human life as a whole; it spreads regardless of any cultural differences or social structures. Secondly, it is irreversible in its impact. Until now, bioethics has produced a corpus of knowledge based on ethical theories that were mainly focused on human agents

1 Moral foundations and actual methods of bioethics are not prepared for the task of understanding this novel way of doing medical science and providing health care (which has also large consequences for future politics and economy).

2 The quotation marks used here want to raise a doubt about the general accepted meaning of this term. I will explain my own use in the third section.

3 An exhaustive survey conducted by the European Union in 2014 established that six out of ten Europeans went online looking for health information: https://ec.europa.eu/digital-single-market/en/news/europeans-becoming-enthusiastic-users-online-health-information (accessed on May 12, 2017).

4 Or, at least, since the concept of "person" was extended to nonhumans, such as dolphins, chimps, bonobos, or elephants.

and living subjects, as one can see in the taxonomy proposed by Arras (2016). New digital technologies swing the focus from the human agent to an efficient treatment of information (for many purposes), something that is part of the posthuman condition. This being the case, the fundamentals of bioethics (e.g., the principles of care and compassion) are at odds with an infocentric doctrine. Moreover, the privilege of human agency in moral decision-making presupposes a different conception of attention, time, and persons than infocentrism, a doctrine based on the assumption that everything can be reduced to data and information. Hence, the question we have to answer is this: is it possible to approach this shift from the "traditional" or "canonical" understanding of agency provided by bioethics? If not, what is to be done in a world of shared agency with artificial kinds?

10.2 The digital epidemiology

Viruses and ideas can spread within a population in a matter of days or weeks. It is enough to have a set of nodes and connections proportionally distributed within that group. In this section, I explore the contemporary *digital setting* and its properties, keeping with the metaphor of contagious viruses. Moreover, I explain why the changes made possible by this setting are radical.

Algorithms, neural networks, digital codes, digital data, and even digital semantic information are not merely objects to be handled in a neutral way. They are encapsulated in computational processes; they act as a technology based on languages and logical operations, able to alter not only information but also human perception, cognition, and beliefs. It is better to think of them as a clique of causal and powerful tools-as-features, and at the same time, as human extensions and environments of our experience (McLuhan 1994). As mediums of our human experience, they also convey the means for understanding it. We act both through them and according to them. The problem of agency is stringent: are they mere functional instruments "in our hands" or are we all together forming a joint apparatus (Introna 2014)?

Hopefully, the term "setting" is not used in an arbitrary manner: digital forces set guidelines for inquiry, knowledge, and affordances for action. They set the modes of our existence, generating not only the places but also the rhythms of experience. Their permeation of the social world is a cold fact – if we acknowledge that >3.5 billion people are online and half of them are on Facebook[5] – and their pace of pervasion is on the rise. In other words, the artificial digital realm is spreading like a virus. However, a biological virus is not bad in itself, as it can occasionally alter our DNA: around 8% of human genetic material originates in viruses, and retroviruses and

5 For other figures in real time, one can access the Internet Live Stats website: http://www.internetlivestats.com (accessed on March 27, 2017).

bornaviruses play a role through endogenization in the reproductive success of our species (Feschotte 2010; Horie et al. 2010) In the same way, a cultural "meme" or "virus" can link human minds for the greater good: new conceptions of freedom, equality, solidarity, and justice are historical facts and their disseminative action was a virus-like process. Still, this exploit could change not only the fundamental structures of life (as bioinformatics has already transformed biology and medicine into information sciences, everything is possible when tinkering with life) but also our intellectual conceptions of it.

In order to determine why the info-computational shift is so substantial and irreversible, we have to take a step back and get a bird's-eye view of the general purposes of these digital technologies. A first characteristic is their capacity to be *transformative*: digital innovation is incremental and cumulative, inasmuch as it frequently changes the forms of perception, mental states, social relations, and general welfare. For example, algorithms not only metamorphose data into new information but also alter the manner in which humans process their increasingly computational environment (Voinea 2016, 592–594).

Transformation does not come alone; it has a companion: *pervasiveness*. Pervasive or ubiquitous computing is not limited to what is visible to persons, as in the case of our desktop computers and laptops. Computation and data processes are now everywhere, from wearables and home appliances to cars, planes, hospitals, and supercomputers. Perhaps the best way to understand this change is to witness the technological subrevolution of smartphones and realize how these tools have transformed all of us into agents in social media. The next such revolution will be accomplished through brain–computer interfaces, especially in medicine (Wolpaw et al. 2002, 786–7).

Secondly, the digital realm is *emergent* and *generative*: its properties are constantly changing, and its aggregation into new formulas is impossible to predict. The idea of generativity within information systems is especially important because it valorizes individual freedom of choice, while also reminding us that we should not completely trust our forecasting capacity. To be generative means to be capable of producing "unanticipated change through unfiltered contributions from broad and varied audiences" (Zittrain 2008, 70). This is exactly what the Internet has done: these technologies can be leveraged because they are versatile enough to meet almost any task. It is easier to use them than to use almost any other tool because of their accessibility (Zittrain 2008, 71–3). All this time, we were not able to predict any of the changes happening within our online empire. We did not guess anything about the rapid rise of Facebook or about hacking politics through "computational propaganda" in 2016 (Woolley and Howard 2016). Expecting the unexpected seems to be the rule of the game.

The third much-admired virtue of information and computer technology is its *effectiveness*. These technologies are the most powerful tools we have ever had for treating huge amounts of data in a limited amount of time. Their success in accompanying humans in economic and social tasks is historically outstanding. This efficacy creates folklores about computers and algorithms being all-problem-solving engines and also raises the hope of liberating humans from their burdens. Indeed, digital tools are more efficient than analogue ones simply because they work with discrete entities, i.e., numerical values that can be perfectly replicated without any loss of information or structure. In fact, if we can convert them to newer formats, these objects are not only placeless, but also timeless. The equivalence of type and token in the case of digital information processing makes them perfect candidates for eternal endurance without the degradation inherent to analogue objects. They are the Cartesian and Leibnizian dream of precision, exactitude, and universality, but there is a weakness. Computation, algorithms, neural networks, software, and digital hardware are prone to errors. Error is a part of the technological condition of these particular artifacts, as well as of any kind of technology. In an era of ubiquitous computing, we cannot always detect deep errors, but only some of the visible glitches. A broken hammer is the image Heidegger (1962, 98) related to and used for explaining the essence of technology, but in our case, the situation is quite different as we cannot even know when the tool is wrecked. Although a digital tool is, in a Heideggerian jargon, "ready-to-hand", it is not always apparent when it "breaks". Resembling fine exploits, errors can pass undetected for a long period. They can also create a cascade of errors that can disrupt, for instance, the activity of an entire hospital within seconds[6]. The opacity of digital objects – even the open source ones – is frightening since they can be socially disruptive.

The digital setting is almost comprehensive and complete in creating the new technological lifeworld, at least for half of the Earth's population. The viruses of digitalized society and the digitization of life – two different manners of the expansion of the digital realm –are spreading fast because they help us survive. The difficulty resides in understanding that we have changed not only the setting for life but also the artificial ambient that we interact and coevolve with. Everything is now computable because it can be expressed as data. In this setting, thinking about the uniqueness of human agency and its mastery over the world and life *without* accepting the causal effectiveness, generativity, and pervasiveness of the artificial kinds is doomed to miss the point, leading to its irrelevance, despite its moral value. To explain how to avoid it, I will sketch an alternative view to the "traditional" assumptions in bioethics related to life, objects, and agency.

---

6 In the making of this article, a case of hospital cyber security lapse emerged in the UK. The National Health Service (NHS) was attacked by the 'WannaCry' ransomware. For more information, follow *The Guardian* link: https://www.theguardian.com/society/2017/may/12/hospitals-across-england-hit-by-large-scale-cyber-attack (accessed on May 13, 2017).

10.3 Ways of understanding the info-computational turn

Three categories could help us decipher the info-computational turn examined above: logical malleability, infocentrism, and secondary agency. These concepts should also inform an alternate view in bioethics. What I have called an inadequate mainstream approach in bioethics (within the digital setting of our technological lifeworld) is shaped by assumptions about normativity, human agency, and universality of moral judgment (Zylinska 2009, 9). Zylinska (2009, 4–11) recognizes the following assumptions within mainstream bioethics: normativity is "predefined" by a specific idea of the good, agency resides solely in human rational subjectivity, and there is continuous demand for universalization of judgments even when it comes to particular cases and their substantial differences. I do not endorse her ensuing critique and positive proposal, but I believe the "diagnosis" is appropriate.

Logical malleability is the main capability that computers have. As James Moor (1985), one of the founding fathers of computer ethics, puts it: "Computers are logically malleable machines in that they can be shaped to do any task that one can design, train, or evolve them to do". Moor (2008, 35) highlights two levels of this capability: the syntactic one – because computers are able to process numbers by a multitude of logical operations – and the semantic one – the capacity to represent "anything we wish" through numerical values and logical states. Without a doubt, semantic malleability is revolutionary in the sense that information processes could be used to represent, model, virtualize, and simulate phenomena, hence producing "a materialization of our conceptual knowledge of the world" (Dodig-Crnkovic 2004, 3). Computers are "universal tools" (Moor 2008, 35), adaptable to tasks that were impossible for other tools. One consequence of this malleability is the incessant novelty and the lack of predictability that accompanies the use of computers and networks. Difficulties produced by the unexpected possibilities of computing are evident in the case of creating information ethics policies. These policies could only approximate our present concerns, so they must continuously adjust to the new problems we encounter (Moor 1999, 68). Thus, it is increasingly difficult to substantiate ethical guidelines or an idea of normativity in an aprioristic manner, beyond actual experience; it rather favors particularism and case-based approaches. To make the matter even more complicated, this generic capability is also present in the case of genetics and biotechnology by making life more malleable (Moor 2008, 35). What we witness today is the convergence of these two kinds of technology: it becomes easy to mediate life phenomena through datafication and computing. In fact, bioinformatics, as an applied and interdisciplinary science, is the best case for this conjunction because it puts computation at work to decipher and produce the "bricks" of life.

Computability *qua* logical malleability is a condition of possibility for infocentrism, and datafication is its majestic method. Datafication is happening worldwide at an exponential growth, similar to its complement, digitization. Unlike digitization, which is a process of converting analog information into binary code, "to datafy a phenomenon is to put it in a quantified format so it can be tabulated and analyzed" (Mayer-Schönberger and Cukier 2013, 78). Datafication is making possible infinite analyses in an "endless stream" (Mayer-Schönberger and Cukier 2013, 84) and anything could be turned into data, which in turn could be aggregated to reveal unexpected patterns and configurations (Mayer-Schönberger and Cukier 2013, 86). In this horizon of endless possibilities, the rise of infocentrism has just been a matter of time.

In a paper published in 1986 about hospital information systems, the term "infocentric" is used to describe the rapid development of health information systems (with the help and involvement of the medical personnel) and how they might be beneficial for the future of health care (O'Desky 1986). It seems that 30 years ago, the problem of responsibility within a plural, almost psychedelic, world was extremely pressing: "Everyone with computer responsibility must be vigilant to the kaleidoscopic modifications in requirements if the infocentric based systems are to be a propitious tool for the suppliers of healthcare services" (O'Desky 1986). Even today, this is sound advice.

According to an infocentric metaphysical doctrine – in ontology and information ethics – the concept of "data entity" is of paramount importance (Floridi and Sanders 2001, 55). It sounds like a benign switch from anthropocentric or biocentric views to a more inclusive perspective; metaphysically, it is opening the way to a world composed and populated by multiple kinds of data entities. The shift is actually subtler, however. If granting differentiated moral status to data entities is not such a radical perspective, conceptualizing human persons as data entities certainly is. The net advantage of the infocentric view resides in its attempt to overcome the belief that only life phenomena are morally worthy. Infocentrism has now become a varied set of beliefs, sometimes inscribed in an ideology or a doctrine, which drive the continuous expansion of the digital technological lifeworld, instead of the macroethical framework Floridi intended it to be. We (started to) believe that anything, from simple to complex and from natural to artificial phenomena, can be construed as a data entity. Within Floridi's macroethics framework, the infocentric understanding of processes, things, and persons is sophisticated and worthy precisely because it addresses a key question: "'What is good for an information entity and the infosphere in general?' The answer is provided by a minimalist theory of deserts: any information entity is recognised to be the centre of some basic ethical claims, which deserve recognition and should help to regulate the implementation of any information process involving it" (Floridi and Sanders 2002, 8). In sum even we, human persons, are information entities; life is a massive stream of logically malleable data, but where are the ethical claims and the deserved moral recognition? One instance of infocentrism could be seen in life sciences, where bioinformatics transforms biology into information science – "contemporary biology works with vast bodies of data that the unaided human mind is incapable of processing effectively" (Griffiths and Stotz 2013, 145) – and reshapes life primitives (e.g., genes) into subjects of datafication.

At this point, we need to delve into another key aspect that could deepen our understanding: shared agency. Agency is perhaps the most significant (and essentially contested) concept in biomedical ethics, mainly because it makes possible the distinction between (autonomous) agents and (dependent) patients. It is also vital to determine agency to delegate responsibility and accountability for choices and actions. Mainstream bioethics identifies autonomy and agency as being something distinctive and uniquely human (O'Neill 2002, 6–7), which is a contestable assumption in highly technologized societies. Could agency be an exclusively human feature in our technological lifeworld? I believe it is not an exclusive feature of humans. I propose to understand agency as relational and shared between humans and different digital artifacts we encounter or rely on in apprehending our experiences.

Secondary and surrogate agency could be two alternative ways to grant artifacts the capacity of choice and goal orientation (Mitcham 2014, 13–22). In Mitcham's taxonomy, secondary agency emphasizes the political nature of artifacts. They could influence society because they are socially normative or nomologically deterministic, but this claim still assumes the ontological distinction between humans and objects. This ontological distinction has been refuted by the Actor Network Theory, another school of sociological thought that proposes that artifacts are proxy agents and are constitutive to actions. From simple calculus done with pocket calculators to high-frequency trading, genetic sequencing, and the International Space Station, actions are made possible, supported, delegated, or guided by digital artifacts. Without a computer, it seems impossible to set and achieve many of our aims. Intentionality, in a strong sense, is deeply connected with the idea of having consciousness, beliefs, and desires but is not a constraint for these categories of agency (Mitcham 2014). Instead, we should admit that many digital artifacts are oriented, have plans, and are purposive and are, thus, intentional (somehow like animals). In the case of computers and algorithms, they exhibit derived intentionality from their designers (Johnson 2006, apud Introna 2014, 34) and act upon the world as such. The more sophisticated the computing method, such as machine learning and deep learning centered on neural networks, is, the more complicated it becomes for us to understand "artificial intelligence" agency[7]. It is not difficult to agree on the fact that at least in two instances, (digital) artifacts exhibit agency: when they prove to be causally successful in combination with humans and when they are directly embedding human agency (Johnson and Noorman 2014, 148–52). Nonetheless, it is essential to not conflate

7 "No one really knows how the most advanced algorithms do what they do. That could be a problem." is a headline of an article published by *MIT Technology Review* on April 11, 2017: https://www.technologyreview.com/s/604087/the-dark-secret-at-the-heart-of-ai/ (accessed on May 13, 2017).

moral agency and sociomaterial agency in the case of artifacts, but the former could be seen as a product of the latter.

The proposed concept of "shared agency" is similar to what Introna (2014, 40) called "co-constitutive agency" in humans and artifacts. Even in medical and health-care settings, the human person is less and less the sole agent. In military or other combative industries, such as chess, go, and financial trading, algorithms will soon replace human intention, choice, and decision as they have already surpassed us in efficiency. The view of a single and unique (sometimes even under divine spell) category of agents – rational human beings using language to express intentions– is at odds with our technological lifeworld, which embraces us and sometimes even takes control of our lives.

These statements should not be read in a deterministic manner, but in a stoic one. How much of human intentions and deeds are exclusively human? And how many of them are shared between humans and their artifacts? It is not our task to count them but to make the second question valid by experiencing the kind of situation it addresses. The need for intelligent assistive technologies in health-care settings for challenging patients (Ienca et al. 2016; 2017) or the sophisticated algorithms that help identify patterns are two instances or types of events wherein we share agency with artificial kinds. These info-computational entities are constitutive to our technological lifeworld, even if they experience nothing of it.

10.4 Conclusions

This excursus was meant to engage the reader in a new way of thinking about vital assumptions in biomedical ethics. For this, I presented an alternative view regarding our digital tools and environments of human experience, which is opposite to the neutrality and functionalist views. Computers, algorithms, neural networks, and data sets are not neutral instruments operating only as a result of our agency, but rather are digital artifacts that present not only affordances but also shape the human thought and action. The liberal–instrumental view should be overcome if we want to understand the realities in health-care settings, wherein agents, patients, and digital artifacts form a joint apparatus (Introna 2014). The causal efficacy in the malleability of life that computers and algorithms display is a strong rationale for changing our moral view and explanatory vocabulary so as to accommodate shared agency. Obviously, this does not imply the conflation of intentionality and subjectivity in making choices but rather the conceptual broadening about how and to whom to designate responsibility in health-care acts. As I claimed before, it is not a deterministic view – we are not trapped in a technological lifeworld without any control or choice –but more of a social constructivist stance by which artifacts have "interpretative flexibility" in interacting with us. Simply put, they are active social constructions, not inert objects around us (Brey 2005, 67–9).

As I stated before, the infocentric doctrine is convergent with a soft social constructivism. When the efficacious treatment of information is at stake, we are compelled to see even human persons as data entities. Remaining just an economic and social ideology, infocentrism could be a threat to our survival if ethical considerations are left out. On the other hand, infocentrism understood as a comprehensive approach to life and to the technological lifeworld can help us mediate between natural and artificial entities. Mediation is a process of integration and adaptation to a complex system made of organisms, information, computational mechanisms, cognition, and semantic meanings. Artificial and natural phenomena are a continuum that could not be divided into different realms. Datafication is both a material process of creating informational representations of human features and activities, as well as an intellectual approach to the complexities of nature and life. Its reductionist method should be balanced by the human social meanings instilled in the process.

The European Data Protection Supervisor, an independent body of the European Union, has a strategy for the ethical dimension *beyond* the legal rules regulating data protection[8], a strategy that is based on an opinion document from 2015 called "Towards a new digital ethics" (European Data Protection Supervisor 2015). I believe this official opinion is a paradigmatic example for the "traditional" view in bioethics, which opposes persons and their agency to artifacts and systems seen as inert functional tools. Liberal–instrumental, as well as being reductionist, this utopian view is driven by the belief in human uniqueness and hence by our ontological and therefore moral dissimilarity. The author(s) of this paper put "human dignity" at the core, thus producing an individualistic and humanistic ethical proposal that is admirable and optimistic, but misguided. A moral upgrade is needed for not only the right to privacy but also the assumptions of this ethical approach. For example, one of the suppositions of this document is that "technology is controlled by humans" (European Data Protection Supervisor 2015, 14) or at least through the decisions we make about its development. While the proposals target real problems – such as the opacity and secrecy around data absorption practices, the need for instillation of privacy values directly into digital design, and so on (European Data Protection Supervisor 2015, 10) – the idea of citizen consent and control over data use as a policy solution is neither complete nor effectual. This grounding paper is an example of aprioristic thinking founded in a deontologist concept, i.e., human dignity, where agency is idealistically solely human. Although it takes into account the pervasiveness of datafication and the emergent powers of algorithms fed by Big Data, it does not construe them in their right: as forming with us a "joint apparatus". In Moor's words, this is an attempt to fill a policy vacuum without clarifying conceptual muddles.

8 More information about the strategy could be found here: https://edps.europa.eu/data-protection/our-work/ethics_en (accessed on 5 June 2017).

Our technological lifeworld has become an info-computational media populated by data and algorithms, an artificial environment for life and shared experiences. The previous sketch of three alternative assumptions for bioethics – it is hardly possible to substantiate ethical guidelines or an idea of normativity in an aprioristic manner; moral status is a function of data entities, not something solely human; agency is plural and thus is shared or sometimes delegated – tried to chart a proposal for a posthuman bioethics. Posthuman is perhaps not the best expression available, but it covers the idea of a shift from a world centered on self-contained and exclusively human agency to a more comprehensive and relational way of thinking. The "posthuman" label should be understood as a rebuttal of biocentrism and anthropocentrism by moving closer to conceptions we encounter in population ethics or in discourse about biosocial and technical systems. Posthuman bioethics is "environmentalist" without losing the humanistic stance. The question regarding how suitable an infocentric bioethics is in practice remains to be settled. The moral principles in bioethics could be reconceived as relying on these new assumptions, in a postindividualistic manner that accepts formal primacy of causal digital artifacts in affording actions in a world of ambient algorithmic intelligence.

References

Arras, John. 2016. "Theory and Bioethics." In *The Stanford Encyclopedia of Philosophy* (Winter 2016 Edition), Edward N. Zalta (ed.). Metaphysics Research Lab, Stanford University. https://plato.stanford.edu/archives/win2016/entries/theory-bioethics/.

Brey, Philip. 2005. "Artifacts as Social Agents." In *Inside the Politics of Technology: Agency and Normativity in the Co-Production of Technology and Society*, edited by Hans Harbers, 61–84. Amsterdam, The Netherlands: Amsterdam University Press.

Dodig-Crnkovic, Gordana. 2004. "On the Importance of Teaching Professional Ethics to Computer Science Students, Computing and Philosophy." In *Computing and Philosophy E-CAP 2004*. Associated International Academic Publishers.

European Data Protection Supervisor. 2015. "Towards a New Digital Ethics. Data, Dignity and Technology." Opinion 4/2015. Brussels: European Data Protection Supervisor. https://edps.europa.eu/data-protection/our-work/publications/opinions/towards-new-digital-ethics-data-dignity-and_en.

Feschotte, Cédric. 2010. "Virology: Bornavirus Enters the Genome." *Nature* 463 (7277): 39–40.

Floridi, Luciano. 2014. *The Fourth Revolution: How the Infosphere Is Reshaping Human Reality*. Oxford University Press.

Floridi, Luciano, and Jeff W. Sanders. 2001. "Artificial Evil and the Foundation of Computer Ethics." *Ethics and Information Technology* 3 (1): 55–66.

Floridi, Luciano, and Jeff W. Sanders. 2002. "Mapping the Foundationalist Debate in Computer Ethics." *Ethics and Information Technology* 4 (1): 1–9.

Griffiths, Paul, and Karola Stotz. 2013. *Genetics and Philosophy: An Introduction*. Cambridge, UK; New York: Cambridge University Press.

Heidegger, Martin. 1962. *Being and Time*. Translated by John Macquarrie and Edward S Robinson. Oxford, UK: Blackwell Publishers.

Horie, Masayuki, Tomoyuki Honda, Yoshiyuki Suzuki, Yuki Kobayashi, Takuji Daito, Tatsuo Oshida, Kazuyoshi Ikuta, Patric Jern, Takashi Gojobori, John M. Coffin, and Keizo Tomonaga. 2010. "Endogenous Non-Retroviral RNA Virus Elements in Mammalian Genomes." *Nature* 463 (7277): 84–87.

Ienca, Marcello, Fabrice Jotterand, Bernice Elger, Maurizio Caon, Alessandro Scoccia Pappagallo, Reto W. Kressig, and Tenzin Wangmo. 2017. "Intelligent Assistive Technology for Alzheimer's Disease and Other Dementias: A Systematic Review." *Journal of Alzheimer's Disease* 56 (4): 1301–40.

Ienca, Marcello, Fabrice Jotterand, Constantin Vică, and Bernice Elger. 2016. "Social and Assistive Robotics in Dementia Care: Ethical Recommendations for Research and Practice." *International Journal of Social Robotics* 8 (4): 565–573.

Ihde, Don. 1990. *Technology and the Lifeworld: From Garden to Earth*. Indiana University Press.

Introna, Lucas D. 2014. "Towards a Post-Human Intra-Actional Account of Socio-Technical Agency (and Morality)." In *The Moral Status of Technical Artefacts*, edited by Peter Kroes and Peter-Paul Verbeek, 31–53. Philosophy of Engineering and Technology 17. Springer Netherlands.

Johnson, Deborah G., and Merel Noorman. 2014. "Artefactual Agency and Artefactual Moral Agency." In *The Moral Status of Technical Artefacts*, edited by Peter Kroes and Peter-Paul Verbeek, 143–58. Philosophy of Engineering and Technology 17. Springer Netherlands.

Mayer-Schönberger, Viktor, and Kenneth Cukier. 2013. *Big Data: A Revolution That Will Transform How We Live, Work, and Think*. Houghton Mifflin Harcourt.

McLuhan, Marshall. 1994. *Understanding Media: The Extensions of Man*. Edited by Lewis H. Lapham. Reprint edition. Cambridge, Mass: The MIT Press.

Mitcham, Carl. 2014. "Agency in Humans and in Artifacts: A Contested Discourse." In *The Moral Status of Technical Artefacts*, edited by Peter Kroes and Peter-Paul Verbeek, 11–29. Philosophy of Engineering and Technology 17. Springer Netherlands.

Moor, James H. 1985. "What Is Computer Ethics?" *Metaphilosophy* 16 (4): 266–75.

Moor, James H. 1999. "Just Consequentialism and Computing." *Ethics and Information Technology* 1 (1): 61–65.

Moor, James H. 2008. "Why We Need Better Ethics for Emerging Technologies." In *Information Technology and Moral Philosophy*, edited by van den Hoven, Jeroen and Weckert, John. Cambridge Studies in Philosophy and Public Policy. Cambridge University Press.

O'Desky, Robert I. 1986. "An Infocentric View of the Hospital Information System." *Healthcare Computing & Communications* 3 (1): 44–46.

O'Neill, Onora. 2002. *Autonomy and Trust in Bioethics*. Cambridge, UK: Cambridge University Press.

Opriş, David, Sebastian Pintea, Azucena García-Palacios, Cristina Botella, Ştefan Szamosközi, and Daniel David. 2012. "Virtual Reality Exposure Therapy in Anxiety Disorders: A Quantitative Meta-Analysis." *Depression and Anxiety* 29 (2): 85–93.

Swan, Melanie. 2013. "The Quantified Self: Fundamental Disruption in Big Data Science and Biological Discovery." *Big Data* 1 (2): 85–99. doi:10.1089/big.2012.0002.

Voinea, Cristina. 2016. "Guvernare fără guvernanţi: politica prin algoritmi şi Big Data." *Revista de filosofie* LXIII (5): 583–95.

Wolpaw, Jonathan R., Niels Birbaumer, Dennis J. McFarland, Gert Pfurtscheller, and Theresa M. Vaughan. 2002. "Brain-Computer Interfaces for Communication and Control." *Clinical Neurophysiology* 113 (6): 767–91.

Woolley, Samuel C., and Philip N. Howard. 2016. "Political Communication, Computational Propaganda, and Autonomous Agents — Introduction." *International Journal of Communication* 10 (0): 9.

Zittrain, Jonathan. 2008. *The Future of the Internet and How to Stop It*. Yale University Press. https://dash.harvard.edu/handle/1/4455262.

Zylinska, Joanna. 2009. *Bioethics in the Age of New Media*. 1st edition. Cambridge, Mass: The MIT Press.

Adriana Paladi, Victoria Federiuc

11 The Principle of Autonomy in Palliative Care: The Moldavian Perspective

11.1 Introduction

Advancing medical technology has made it possible for many people to live longer than ever before. However, when the possibility of curing patients no longer exists, care should not be discontinued, but rather should shift to comforting the patients and their families. This type of care, labeled palliative care, is defined by the World Health Organization (WHO) as "an approach that improves the quality of life of patients and their families facing the problems associated with life-threatening illness, through the prevention and relief of suffering by means of early identification and impeccable assessment and treatment of pain and other problems, physical, psychosocial and spiritual."[1]

Palliative care aims to improve the quality of life of patients. Good quality of life is usually related with control of symptoms and avoidance of treatment side effects, but it is more than that. While defining this concept, it is necessary to encompass the individual's (patient) perspective on biological comfort, social and psychological well-being, as well as spiritual integrity. Thus, any palliative intervention requires a focus on the person whose quality of life is at stake, because the description of quality of life is best made "in individual terms, and depends on present lifestyle, past experience, hopes for the future, dreams and ambitions. A good quality of life can be said to be present when the hopes of an individual are matched and fulfilled by experience." (Calmen 1984, 124–5).

In other words, quality of life designates the subjective perception of one's life, measured in the context of personal experience and expectations, as there is no obvious general standard for it. Two different patients could have entirely different views of the best quality of life, even if they have the same illness. Thus, the satisfaction or contentment of patients in terms of end-of-life care (among others) will depend on how their personal expectations are fulfilled (Jackson et al. 2001).

Accordingly, palliative care involves knowing the patient's personal priorities, encouraging the patient's realistic goals (appropriate to the medical condition), and helping the patient to reach these goals via physical, psychological, social, and spiritual interventions (Rummans and Bostwick 2000; Pantilat et al. 2008).

1 WHO definition of palliative care available on: http://www.who.int/cancer/palliative/definition/en/

Focusing on personal perspective while establishing the measure of the best quality of life for each patient suggests a strong interconnection between palliative care priority and the principle of autonomy. Respect for autonomy emphasizes the idea of individual decision-making. Shared medical decisions allow the patient to maintain control over his/her own life, strengthening the patient's perception of being treated with dignity. The beneficial outcomes result from the patients' involvement in health-care decision-making (Epstein and Street 2000).

Presuming to know without asking what quality of life is desired by a patient is morally wrong. It violates the patient's autonomy and, consequently, is a form of disrespect toward the dignity of the patient. Ignorance of the patient's perspective may lead to actions or omissions that do not meet the patient's needs and goals and may even be harmful for them.

Consequently, it seems obvious that the respect for individual autonomy is an indispensable condition for achieving the priority of palliative care – namely, enhancing patient's quality of life.

Nevertheless, there are scholars who underline the local character of the principle of autonomy. For instance, Daniel Fu-Chang Tsai documents that the emphasis of Western medical ethics on autonomous decision-making is not shared by all cultures (Fu-Chang Tsai 2008). The individualistic approach to autonomy is a Euro-American value and cannot be ethically applied in all settings (Sargent and Smith-Morris 2001).

In this article, we are not going to challenge the plausibility of this anthropological claim; our intention is to argue that the disrespect for individual autonomy in palliative settings, in a country with strong paternalistic and communitarian traditions as Moldova, could result in health professional's incapacity to act in the patient's best interest, similar to countries with old liberal traditions.

To that extent, we will explore the reasons and effects of the lack of honest communication and noninvolvement in decision-making on patient's quality of life, challenging the assumptions that founded the physicians' decision of hiding incurable diagnoses. Furthermore, a model, included recently in Moldavian health-care professionals' education, of honest sensitive communication with patients and some of its advantages are presented.

11.2 Principle of autonomy

Individual autonomy usually refers to the capacity to live one's life according to reasons, values, preferences, and interests that are taken as one's own and not distorted by external forces. Beauchamp and Childress highlight that at least two conditions are essential for autonomy: liberty (independence from controlling influences) and agency (capacity for intentional action). Consequently, at the minimum, personal autonomy encompasses self-rule (self-governance) that is free from both controlling influence by others and limitations that prevent meaningful choice, such as inadequate understanding (Beauchamp and Childress 2013, 101).

Applied to palliative care, the principle of autonomy requires that patients' wishes be respected and that they be helped to participate in decisions about their care, to the extent that they are willing and able. When patients are unwilling to or incapable of making their own medical decisions, this principle requires that the decisions taken by a patient's legal representative be based on the patient's earlier wishes.

More particularly, the requirement of patient involvement in health-care decisions obliges physicians to act as partners by maintaining open and frequent communication about a person's health condition, available care options, and prognosis to ensure the person or their representative fully understands the information. Clear information enables the patient to make decisions in accordance with his/her beliefs, values, and interests, and the physician is obliged to respect the patient's informed decision, even if these differ from the family or the health-care professional's views. Respect for patient decision leads to long-term trust between physician and patient, increasing patient compliance to treatment, enhancing outcomes, and elevating mutual satisfaction. In other words, autonomy grounds the model of the doctor–patient relationship in which "patients are thought of as equal partners in their treatment, in which treatment is given only with the informed consent of patients, in which patient satisfaction is an important indicator of professional adequacy". In this model of doctor–patient relationship, autonomy is seen as a precondition of genuine trust (O'Neill 2015, 18–9).

11.3 Moral requirement of patient information

As mentioned, the principle of autonomy places a responsibility on the doctor to ensure that his/her patient is fully informed. Complying with this requirement raises practical difficulties when a patient diagnosed with life-threatening illness is considered. Even now, many physicians are not ready to break bad news to patients[2].

Cases in which physicians avoid open communication with patients that have life-threatening medical condition are not uncommon. Vladimir Poroch and Andrada Pârvu (2013) point out that many physicians mistakenly hope that someone else will inform the patient about life-threatening conditions (another colleague). As this never happens, patients are frequently left ignorant.

An empirical study on 137 patients with unfavorable diagnoses (patients with chronic obstructive pulmonary disease [COPD], acquired immune deficiency

---

2 The concept of bad news is associated in this article mostly with information about terminal diagnosis, even if, generally, it is understood as "any news that drastically and negatively alters the patient's view of her or his future" (Beauchamp and Childress 2013).

syndrome [AIDS], and cancer), family members, health-care workers, and physicians reveals that honest communication is seen by all participants as an important aspect of palliative practice. This study emphasizes, at the same time, the patients' unmet need of information about their diagnosis, witnessing indirectly the lack of honest communication in palliative care practice. In the study, a patient with COPD confessed, "Dr ____ has never told me my illness was serious. I've asked him, but he doesn't answer." (Wenrich et al. 2001, 869). Many health professionals consider, in principle, that patients have to be informed about their diagnosis, but only few of them disclosed bad news in practice (Grassi et al. 2000).

11.4 Some reasons for covering up bad news

Physicians may find it difficult to discuss issues around end-of-life situations because of the strong attachment with the dying patient and family developed during the long medical practice. Diverse forms of concerns also may preclude physicians from communicating openly with dying patients, such as fear to confront their own mortality and fear of displays of emotions (i.e., fear of emotional outbursts, fear of experiencing significant stress, fear of being blamed as a messenger of sad news, fear of not being prepared to manage the patient's emotional reactions, fear of iatrogenic complications in the patient's situation) (Anderson 2000; Ptacek et al. 1999).

Communication with dying patients is affected by cultural differences as well. Many cultures do not support the idea of full disclosure of bad diagnoses or prognoses, while others require disclosure to family members or community leaders. Health-care professionals outside the Western cultural tradition often conceal serious diagnoses from patients, because of four primary reasons: 1) discussion of serious illness and death is an expression of disrespect for the belief that individual destiny is determined by God (Filipino culture); 2) open communication of serious illness is inhuman because it may provoke unnecessary depression or anxiety in the patient (Chinese culture); 3) physicians are expected to maintain patients' optimism and direct disclosure may eliminate it (Bosnian culture); 4) speaking aloud about a condition makes death real because of the power of spoken words (Navajo) (Searight and Gafford 2005).

All these cultural reasons (except the last one) are reflected in the paternalistic approaches to physician–patient relationship, which places a higher value on beneficence and nonmaleficence relative to autonomy. Within this approach, physicians are allowed to interfere with patients' wishes if the interference brings greater benefits or prevents serious harms. Professional competence gives physicians authority to decide on behalf of the patient what is better or worse for them.

Paternalism has a long historical tradition and deep roots in collective thinking and perceptions of doctor–patient relations. Nevertheless, in the past decades, globalization of conventional medicine, wide acceptance of this Western legal requirement, and the increase in patients' awareness of their rights have weakened its position. However, the process of transition from paternalistic to nonpaternalistic approaches to practicing medicine is very slow. Changing mentalities is very difficult and, many times, occurs together with replacement of generations.

11.5 The case of the Republic of Moldova

The Republic of Moldova is a country with continuing strong paternalistic and communitarian tradition. In 2005, the Parliament of the Republic enacted the Law of Health No. 263-XVI, Article 11(5), which emphasizes the following: "Medical information on the health status (condition) of the patient and the proposed medical intervention, including the risks and potential benefits of each procedure, the possible effects of refusal of treatment, alternatives, prognosis and other medical information, must be presented by health services providers in a clear language, respectfully and accessible to the patient, minimizing professional terminology". After 10 years, despite being granted by law to have access to information, many patients with bad diagnoses are not adequately informed. For example, a recent study conducted on 228 patients in the chemotherapy ward of the Oncology Institute from Chisinau (Moldova) reveals that only 62.1% of those surveyed knew their cancer diagnosis. The remaining patients did not know about their life-threatening illness. These patients had general information about carcinomas, polyps, and cysts but did not really understand their conditions (Clipca 2016).

There are no published empirical studies about the causes of hiding the diagnoses from incurable patients in Moldova; however, the reason for the tendency to hide serious diagnoses, most fervently invoked by Moldavian physicians in private or formal discussions, is the unwillingness to harm the patient. This reason is based on assumptions that telling the complete truth about a terminal diagnosis a) could have a negative impact on patients' physical, psychological, and spiritual state; and/or b) could take away all hope and optimism of patient. Are these assumptions justified? Does communication about lethal diagnosis have a negative impact on patient well-being? In order to answer this question, let us undertake a consequentialist analysis of a clinical case common in Moldavian health-care practice.

Mrs. R, a 39-year old mother of two children, is admitted into hospital with abdominal pain. The medical examination shows that she is suffering from an incurable pancreatic cancer, with a few months to live and with great suffering before her. Although the woman has specifically asked to be told exactly what she was facing, the physician informs her that the results of the examination are not conclusive and that he will see her weekly as an outpatient. Instead, the physician informed her husband, privately, about her diagnosis – this is a traditionally assumed practice. In Moldova, the family is very involved in the process of caring for a sick member and usually takes care of dying relatives.

Whether the culturally supported physician's decision is beneficial or harmful for the woman and her family is being established through the consequences that would result from it.

One favorable consequence of not telling the truth is that the patient could feel better immediately after the appointment. Nevertheless, improvements in well-being are likely to be short term because pain persistence could rapidly burn it down. Alternatively, painful existence may lead the patient to doubt the professional competence and honesty of the physician. Uncertainty felt by the patient concerning her health condition may generate stress, anxiety, and depression. Lack of information withdraws her chance for coping and planning the future, a situation that could have negative consequences on her entire family. If kept far from the truth, the patient would not be able to make an adequate decision, decisions that would serve her best interests. At the same time, an undue burden is put on the husband, who almost obviously would not be able to deal appropriately with the information acquired, along with handling the stress and anxiety resulting from it.

Therefore, deception leads to much more unfavorable consequences than favorable consequences, and the physician's decision is not in the benefit of the patient and her family; even vice versa is true. Thus, the assumption about the negative impact of disclosing information is unwarranted in this case and in many similar other cases.

Another important question is whether information about lethal diagnosis takes away the patient's hope? An inquiry into this phenomenon is required to answer the above question.

Even if it could be defined in many ways, the concept of hope designates at least the individual's belief that his/her needs, wishes, or goals are being respected. Individual hope is quite dynamic, it is changing all the time because the individual's needs and goals are as well, reflecting the mutability of life circumstances. For instance, at the time of life-threatening diagnosis, the patient may hope for survival, but at the end of the illness trajectory, he/she may hope to not be ignored (Chichinov 2006).

Sullivan (2003) claims that hope at the end of life can come in various forms: hope for comfort, dignity, intimacy, or salvation. On the other hand, hopelessness is not simply the absence of hope, but attachment to a form of hope that is lost. In the palliative context, as well as in any other health-care contexts, patients may lose hope but also may find another hope when they are able to explore positive possibilities of the new circumstances of life and break their attachment to the lost forms of hope.

In palliative settings, when the hope for cure is lost, the primary task of caregivers is to redirect the hope of both the patient and the family toward something more realistic, such as hoping to have well-managed symptoms, to feel less pain in order to enjoy spending time with family and friends, and to accomplish any goals they have before their death. Lowey Susan suggests several palliative interventions that could foster patient and family hope, despite a fatal prognosis: keeping symptoms well managed, encouraging involvement in positive experiences that transcend the patient's current situation, fostering spiritual processes and finding meaning, promoting reconciliation with others (positive personal relationship), assisting patient in setting realistic goals, and focusing the patient's attention on the short-term future (Lowey 2015).

All these interventions applied efficiently in end-of-life care show that it is incorrect to think fatalistically about end of life. Many things can be done to guide hope and increase the quality of life.

Accordingly, the assumption that disclosure of the terminal nature of their illness deprives patients of hope is also mistaken. Even more, there is good evidence that honest communication about health conditions enhances, rather than diminishes, hope because accurate information enables patients to feel empowered about their care and decision-making at the end of life (Davison and Simpson 2006).

If the assumption about the negative impact of honest communication on hope is wrong, it means that those Moldavian physicians who avoid informing their patient about an incurable diagnosis might not be aware that patients with lethal diagnosis or those who are imminently dying continue to hope. In this sense, Ivana Clipca (2016) suggests that physicians in Moldova do not have a clear understanding of the concept of hope, treating it just in terms of survival.

The case analysis, as well as the reflection/thought on the dynamic nature of hope, makes it possible to assert that dishonest communication with patients prevents the Moldavian physician from accomplishing his duty to act in the patient's best interest, i.e., improvement of quality of life. This supposition can be assessed considering empirical evidence. In 2011, with the support of the National Commission of Republic of Moldova for United Nations Educational, Scientific and Cultural Organization (UNESCO), a joint team of professionals from Great Britain and the Republic of Moldova published a study about the personal experiences of 102 families across the Republic of Moldova, families that have cared for their dying loved ones. The study concludes that "dying people suffer significantly, pain is often uncontrolled, while anger, isolation, depression, grief, and fear are often unaddressed by health service" (Kellehear et al. 2011, 95).

Even though this study did not aim to evaluate the gaps in communicating with dying patients, it allows us to think that one reason, among many others, for the great suffering of dying patients in Moldova is the lack of honest communication. It badly affects the dying patient's quality of life; particularly because in the condition of dishonest communication, the health-care professional cannot explore the patient's needs and preferences for end of life and, consequently, is not able meet them.

There are cases when telling the truth about the diagnosis and prognosis could be harmful for a patient's health. If the physician has good evidence for expected harm, then he/she also has the moral right to withhold that information. Nevertheless, this course of action should be carefully considered. In many other cases, as we have argued, the projected harm does not justify overriding the principle of autonomy.

11.6 Delivering bad news in a sensitive manner

Concealment of the truth excludes the patient from the decision-making process and is a violation of a patient's constitutional[3] right to information and self-determination. At the same time, it is also a major obstacle to both fulfilling the patient's needs and increasing the patient's quality of life in Moldova. However, informing patients about the serious health problems should be done with tact and sensitivity.

When the information is delivered incorrectly, it could lead to undesirable consequences for patients and their families. Good communication skills, as well as high competence in delivering bad news, could be significant for avoidance of such consequences.

Learning general communication skills can enable physicians to break bad news in a manner that is less uncomfortable for them and more satisfying for patients and their families. A physician's attitude and communication skills have a crucial role in how well patients cope with receiving bad news (Vandekieft 2001).

Nowadays, there are many guidelines, protocols, and recommendations developed to help physicians deliver bad news. Many of them are adapted from a six-step protocol proposed by Robert Buckman (1984). Following this protocol's steps, individualizing their manner of breaking bad news in accordance with the patient's needs, physicians achieve four essential goals. The first one is gathering information from the patient that allows them to determine the patient's knowledge, expectations, and readiness to hear the bad news. The second goal is providing intelligible information in accordance with the patient's needs and desires. The third goal is offering emphatic support to the patient in order to reduce the negative emotional impact and to prevent exacerbated feelings of isolation. The final goal is to develop a strategy in the form of a treatment plan with the input from and cooperation of the patient, a plan that best meets the patient's needs (Baile et al. 2000).

In 2016, in the State Medical and Pharmaceutical University in Moldova, "*Nicolae Testemițanu*" – the integrated course of palliative medicine – was introduced, including a module on breaking bad news using a six-step protocol. Simulation with standardized patients on the graduate level and role-playing approaches on the postgraduate and vocational levels are used as the teaching methods. One year after the introduction of the course, some preliminary conclusions can be drawn. Thus, graduate students (the fifth-year medical students), as well as postgraduate trainees (family physicians mostly), easily follow the steps of the protocol (Emanuel, von Gunten, and Ferris 1999), which require the following:

1. creating an environment conducive to effective communication and confirming the medical facts of the case (Step 1);

3 Law of Health No. 263-XVI, Article 11(5) of the Republic of Moldova enacts this patient's right.

2. using the open-ended questions about what the patient and family know, understand, and feel about the patient's health (Step 2);
3. finding out what and how much the patient wants to know and, deciding on who should be designated to make decisions on the patient's behalf when the patient prefers not to receive critical information (Step 3);
4. giving the patient and family reassurances that they are not being abandoned through planning for the next steps (i.e., arranges for appropriate referrals, prescribes additional treatment, discusses about potential needs and sources of emotional and practical support, and establishes a time for a follow-up appointment) (Step 6).

However, they face difficulties in offering the information in a sensitive, but straightforward, manner; they fail often in avoiding both a steady monologue and medical/technical jargon, and checking periodically the patient comprehension (Step 4).

Moreover, trainees encounter difficulties in responding emphatically to the emotional reaction of standardized patients (Step 5). They often are not able to acknowledge the patient's emotions, to validate their legitimacy, and to give support to patients through a broad range of their reactions. The lack of emphatic competence instance among our trainees resulted in ignoring the patient's emotions through not allowing time for expression of his/her immediate feelings or in rushing the situation through by quickly starting the next step of the protocol, namely, "Summarizing and planning".

Thus, breaking the bad news in a straightforward and emphatic manner seems to be a very difficult task for many trainees. Moreover, if some of them hesitate to accomplish it during the educational process, they almost certainly will avoid performing it in a clinical setting. This state of reality reveals the necessity to empirically explore the deep cognitive, emotional, and cultural reasons behind this behavior and to construct the educational process on delivering bad news in accordance with the acquired findings to suit the context better.

11.7 Conclusion

The principle of autonomy in the context of providing medical treatment is considered to be paramount in liberal countries and is becoming increasingly significant in countries with communitarian traditions. It requires that physicians have open and honest communication with patients about diagnosis and prognosis to enable patient decision-making. Honest conversation, sensitively and competently navigated, results in a lot of advantages. It strengthens the physician–patient relationship and fosters collaboration among the patient, family, physicians, and other professionals. It allows the patient to express his/her wishes and needs, making them known and able to be met. Honest and sensitive delivery of information enhances the patient's capability to make informed decisions about his/her care, as well as to maintain their personal dignity. At the same time, it allows patients to prioritize and prepare for the future and reduces bereavement suffering for those left behind. Open communication about health condition helps patients to harmonize their expectations to the objective reality (to the real state of affairs), to set realistic goals, and do the best to achieve them. Consequently, all these may finally increase the patient's quality of life.

Providing full information to the patient about his/her health condition when it concerns incurable diseases is a difficult task avoided by many Moldavian physicians, usually on the ground of suspicion that true information may inflict harm to the patient. However, there is evidence that the opposite is true. Dishonest communication precludes Moldavian physicians from knowing the patients' needs and to meet them in a palliative setting, which results in great harm for the patients. At the same time, appropriate disclosure of bad information by the physician requires communication skills and competence developed during special training, which has been inaccessible, actually, for Moldavian physicians until recently.

We are confident that culturally sensitive education of Moldavian physicians in this field will make them aware about the significance of the "principle of autonomy" and will increase their willingness to apply it in palliative practice for the benefit of the patient.[4]

References

Anderson, Ian. 2000. *Program in End-of-Life Care. Communication with Patients and Families.* University of Toronto.

Baile, Walter F., Buckman Robert, Lenzi Renato, Glober Gary, Beale Estela A., and Kudelkab Andrzej P. 2000. "SPIKES—A Six-Step Protocol for Delivering Bad News: Application to the Patient with Cancer." *The Oncologist* 5(4):302-311.

Baker, R., A.W. Wu, J.M. Teno, B. Kreling, A.M. Damiano, H.R. Rubin, M.J. Roach, N.S. Wenger, R.S. Phillips, N.A. Desbiens, A.F. Connors Jr, W. Knaus, and J. Lynn. 2000. "Family satisfaction with end-of-life care in seriously ill hospitalized adults." *Journal of the American Geriatrics Society.* 48(5, Suppl.):61–69

Beauchamp, Tom L., and James F. Childress. 2013. *Principles of Biomedical Ethics.* Seventh Edition, New York: Oxford University Press.

Buckman, Robert. 1984. "Breaking bad news: why is it so difficult?" *BMJ* 288:1597-9.

Calmen, Kenneth C. 1984. "Quality of life in cancer patients – a hypothesis." *Journal of Medical Ethics* 10:124-127.

Chochinov, Harvey Max. 2006. "Dying, Dignity, and New Horizons in Palliative End-of-Life Care." *CA: A Cancer Journal for Clinicians. 56:84–103.* http://CAonline.AmCancerSoc.org

Clipca, Ivana. 2016. *Particularități ale comunicării cu pacienții oncologici.* (unpublished study).

4 We would like to thank Emilian Mihailov and Tenzin Wangmo for their comments and suggestions in editing this article.

Davison, Sara N., and Christy Simpson. 2006. "Hope and advance care planning in patients with end stage renal disease: qualitative interview study." *BMJ* 333:886.

Emanuel, Linda L., Charles F. von Gunten, and Frank D. Ferris. 1999. *The Education for Physicians on End-of-life Care curriculum.* The Robert Wood Johnson Foundation. Accessed April 1, 2017. http://www.euro.who.int/document/e82933.pdf

Epstein, Ronald M., and Richard L. Street, Jr. 2007. "Patient-centered communication in cancer care: Promoting healing and reducing suffering." National Cancer Institute. NIH Publication No. 07-6225.

Fu-Chang Tsai, Daniel. 2008. "Personhood and Autonomy in Multicultural Health Care Settings". *AMA Journal of Ethics.* Volume 10, Number 3: 171-176.

Grassi, L., T. Giraldi, E.G. Messina, K. Magnani, E. Valle, and G. Cartei. 2000. "Physicians' attitudes to and problems with truth-telling to cancer patients." *Support Care Cancer.* 8(1):40-5.

Jackson, J.L., Chamberlin J., Kroenke K. 2001. "Predictors of patient satisfaction." *Social Science and Medicine.* 52(4):609-20.

Kellehear, Allan, Vadim Pogoneţ, Rodica Mîndruţa-Stratan, and Victor Gorelco. 2011. *Îngrijirea bolnavilor muribunzi în Republica Moldova.* Comisia Naţională a Republicii Moldova pentru UNESCO.

Lowey, Susan E. 2015. *Nursing Care at the End of Life. What Every Clinician Should Know.* Open SUNY textbooks. Accessed April 1, 2017. http://pressbooks.opensuny.org/nursingcare/

O'Neill, Onora. 2015. *Autonomy and trust in bioethics.* Cambridge: Cambridge University Press.

Pantilat, S.Z., Isaac M. 2008. "End-of-life care for the hospitalized patient." *The Medical Clinics of North America.* 92(2):349-70, viii-ix.

Poroch, Vladimir, and Andrada Pârvu. 2013. "Îngrijirile paliative." In *Dileme etice la finalul vieţii*, edited by Beatrice Ioan and Vasile Astărăstoae. Iaşi: Editura Polirom.

Ptacek, J.T., Fries E.A., Eberhardt T.L., Ptacek J.J. 1999. "Breaking bad news to patients: physicians' perceptions of the process." *Support Care Cancer.* 7(3):113-20.

Rummans, T.A., Bostwick J.M., Clark M.M., Mayo Clinic Cancer Center Quality of Life Working Group.2000. "Maintaining quality of life at the end of life." *Mayo Clinic Proceedings.* 75(12):1305-10.

Sargent, Carolyn, and Smith-Morris Carolyn. 2012. "Questioning Our Principles: Anthropological Contributions to Ethical Dilemmas in Clinical Practice". *Occasional Papers*.4. http://scholar.smu.edu/centers_maguireethics_occasional/4.

Searight, H. Russell, and Jennifer Gafford. 2005. "Cultural diversity at the end of life: issues and guidelines for family physicians." *American Family Physician* 71: 515–22.

Sullivan, Mark D. 2003. "Hope and hopelessness at the end of life." *American Journal of Geriatric Psychiatry* 11(4): 393-405.

Vandekieft, Gregg K. 2001. "Breaking Bad News." *American Family Physician* 64: 1975-8.

Wenrich, Marjorie D., Curtis J.R., Shannon S.E., et al. 2001. "Communicating with dying patients within the spectrum of medical care from terminal diagnosis to death." *Archives of Internal Medicine* 161(6): 868-874.

Laurențiu Staicu, Octavian Buda

12 Philosophical Foundations and the Role of Counseling in the Ethics of Informed Consent

12.1 Introduction

The postwar outcry against Nazi medical experiments conducted on human subjects led to heated debates regarding the ethical principles of informed consent. In the late fifties, medical and research institutes came to acknowledge the patient's right to decide on subjecting his/her body to high-risk treatments and experiments. However, informed consent and the ethical principles underlying its use and interpretation in clinical cases were given due consideration only much later. In 1972, Jay Katz released a monumental work dedicated to the history of informed consent and the medical process of seeking patients' informed consent (Katz 1972). Although Katz tried to rationally reconstruct the history of informed consent, his steadfast mistrust in the way in which medical and research institutions obtained patients' consent was either criticized or ignored by the medical community. A decade later, in 1982, Martin S. Pernik published a study entitled "The Patient's Role in Medical Decision Making: A Social History of Informed Consent in Medical Therapy" (Pernick 1982). Pernik's study was well received due to his liberal approach and extremely generous definition of the concept of informed consent. Finally, in 1986, Ruth R. Faden and Tom L. Beauchamp put out *A History and Theory of Informed Consent* (Faden and Beauchamp 1986), which was considered an attempt to find an adequate definition for the concept of informed consent and to offer a systematic analysis of its ethical principles.

These important studies became methodological guidelines due to their thorough and spectacular reviews of the thorny issues related to informed consent in the medical practice[1]; nonetheless, they failed to provide a clear-cut account of either informed consent or the ethical principles underlying it. Moreover, after reviewing the recent literature, we arrived at the conclusion that in the past years, there has hardly been any substantial contribution to a better understanding of informed consent, besides refining or updating existing definitions by resorting to comparative analyses and various case studies[2].

This study argues that divergent approaches to informed consent are, to a large extent, determined by previously established philosophical presuppositions. We

1 Grzegorz Mazur, for instance, describes them as foundational or paradigmatic theories, which truly established the main methodological guidelines in the area of informed consent (2012, 1–8).

2 Refer, for instance, Manson and O'Neill (2007) and Hammond (2016).

try to reveal these presuppositions by analyzing the most prominent definitions of informed consent and its underlying ethical principles. In the final section, we put forward an original proposal in an attempt to iron out existing differences, and we suggest the addition of a new intermediary step in the process of obtaining informed consent, one that could lead to a more robust ethical foundation for this sensible and difficult process.

12.2 Informed consent as an expression of liberty

What is informed consent? If we accept the supposition made by Faden and Beauchamp, according to which the informed consent doctrine originates in the care to protect and facilitate clinical patients' and subjects' autonomy and free choice (1986, 235), we need to define consent by resorting to terms such as autonomy or, to be more specific, autonomous action. In other words, informed consent is the free choice of an agent (patient/subject) to accept a specific treatment or surgical intervention.

This brief definition contains three implicit premises: first, that the respective action is meant to improve the patient's health; secondly, that this is going to be a potentially dangerous intervention that may lead to an improvement of the patient's health, but which may also involve risks; thirdly, that there is no other intervention capable of producing the same effects without adjacent risks for the patient's health.

Let us suppose we are dealing with a patient facing an end-stage disease. Doctors recommend an experimental treatment with a good track record in tackling the respective disease, but which may lead to health problems given the patient's condition. Under the circumstances, doctors will ask for the patient's consent or, if the patient is not mentally competent or unable to provide consent, they will seek the consent of the patient's nearest relative.

When the patient is asked to give his/her consent, he/she is provided with all relevant data – both benefits and risks – and procedures concerning the medical intervention. In this scenario, seeking the patient's consent means actually passing the responsibility for the proposed treatment onto his/her shoulders, provided the patient is correctly informed that there is no risk-free alternative.

Shifting the burden of responsibility sends us directly to the idea of autonomy, because the patient is seen as an autonomous person who has the right to decide on any medical procedure that may affect his/her health or endanger his/her life. In other words, informed consent is based on the fundamental premise that a patient or subject is a person possessing freedom of choice over their life or well-being. The choice is an autonomous action of the patient, who freely accepts a procedure involving risks. A doctor may obviously know a lot more than his/her patient in assessing a situation and applying a certain procedure, but that does not give him/her the right to act against the patient's free choice.

There is plenty of room for skeptical worries concerning the veracity and reality of this perspective. More often than not, a physician treats his/her patient not as an autonomous subject, but rather as a biological organism that needs "repairing", or as an individual who has no idea what is best for him. For instance, Katz severely criticizes doctors for encouraging patients to relinquish their autonomy. In his opinion, the most frequent situation is that of patient's manipulation and persuasion by the practitioners, and not that of keeping the relevant information from the patient with the purpose of obtaining the latter's informed consent: "the history of the physician–patient relationship from ancient times to the present (...) bears testimony to physicians' inattention to the right and need of their patients to make their own decisions. Little appreciation of disclosure and consent can be discerned in this history, except negatively, in the emphasis on the inability of patients to grasp the mysteries of medicine and therefore to share the burdens of decision with their doctors" (Katz 2002, 28).

But such doubts are not enough to put an end to a thorough analysis of the definition of informed consent, or even to make it look utopian. Let us suppose that we accept Katz's skepticism and his conclusion that in the real world, doctors ignore the principle of autonomy and patients' free choice. This does not imply that medical practitioners are *necessarily* bound to act this way and to ignore the patient's choices. They are free to act otherwise. For example, they could have the opposite attitude, taking into consideration the patient's choices, and this attitude would be something we would certainly want to enforce.

After all, from a more general point of view, the immorality of human behavior, which is, indeed, real, does not affect or erode the legitimacy of moral rules that all individuals should respect. On the contrary, the more immoral individuals are, the greater is the need to make them consider the governing moral rules as well as the measures required to overcome and reduce their immorality.

Now, going back to the principle of autonomy as an ethical foundation for informed consent, sometimes, this principle seems to be taken as being in opposition with the principle of beneficence concerning the doctor–patient relationship. For example, Faden and Beauchamp consider those principles as the ethical source for two different, even contradictory, types or models of doctor–patient relationship, which were competing for supremacy throughout the history of medicine.

In the model based on the principle of beneficence, one that could be derived from the Hippocratic Oath, the physician carries full responsibility for any medical procedure. The well-being of a patient prevails over his/her freedom of choice and, if need be, it allows the doctor to ignore the patient's decisions. The same principle says that no matter how well informed the patient may be, the final decision belongs to the doctor, given, of course, his/her medical expertise and knowledge.

On the other hand, in the model based on the principle of autonomy, which, according to Faden and Beauchamp, has emerged only in modern times, the fundamental idea is that the patient is the sole agent in charge of his/her fate. The doctor is not a completely autonomous decision-maker, but merely a person who carries out the patient's decision, a sort of delegated specialist acting in a framework set by the autonomous individual – the patient, whom he/she treats according to his/her relevant expertise. Therefore, the principle of autonomy enforces the idea that "the autonomous person is not bound by controlling constraints and is in control of personal affairs" (Faden and Beauchamp 1986, 8).

Even though the tension between the two principles is intuitive and anyone could imagine scenarios in which those principles determine conflicting doctor–patient relations, this does not imply that we are somehow constrained to accept it. Although Faden and Beauchamp consider the opposition between these two principles as essential in any rational reconstruction of the history of medicine, quite a few critics have argued that these principles may collaborate rather harmoniously, based on a deeper presupposition seen as a cause or a foundation for both of them. More precisely, the opposition or the tension between the two principles can be real *de facto*, but hardly so *de jure*. Mazur, for example, argues that although those principles can outweigh each other in different stages of medical practice, this does not imply that they are necessarily at odds with each other. Moreover, "depending on the moral character of both practitioner and patient, they might be used in such a way as to complement each other" (Mazur 2012, 6). Though ambiguous[3], this collaboration of the two principles according to the moral attitudes of practitioner and patient shows that the doctor's expertise is not always challenging (or in opposition with) the patient's choice. Therefore, we do not agree that the two principles determine, necessarily, opposing doctor–patient relationship patterns.

On the other hand, even if we do agree with the possibility of a harmonious collaboration between the two principles in medical practice, we should keep in mind the claim by Faden and Beauchamp that the principle of autonomy has started to gain ground only in modern times, more precisely a time when personal autonomy and freedom have become acknowledged as fundamental human rights, regardless of the context (or clinical situation). Hence, even if they could collaborate, the two principles might as well come to loggerheads, depending on the metaphysical presuppositions that lie deeper than the moral profile to which Mazur refers.

---

3 In order to avoid a wrongful misinterpretation of Mazur's opinion, we must stress that the ambiguity in question refers only to his discussion concerning the theory advanced by Faden and Beauchamp. In the ensuing chapters, the author elaborates on this idea and tries to explain how the moral profiles of both doctor and patient could support the collaboration between the two principles. However, we shall dwell longer on this subject in the last section.

12.3 Informed consent as an expression of morality

The principle of autonomy is widely accepted as being an important foundation in the process of obtaining informed consent, but the principle's preeminence in clinical theory and practice is not considered a natural or necessary consequence, as someone could expect. Moreover, it could be argued that the principle of autonomy by itself is insufficient as a proper moral foundation for the practice of informed consent. For example, it could be argued that the unconditional respect for a patient's autonomy as a person could have immoral results or consequences. For instance, if a patient rejects a clinical intervention and the doctor accepts the patient's decision relying on the respect for the person's autonomy and free choice, the situation may endanger the patient's well-being and even life. In such a scenario, we may wonder whether accepting a person's autonomy is a moral gesture. Perhaps, we may need additional rules to avoid, as much as possible, putting a patient's life at risk.

In other words, placing the autonomy principle as an unconditional and sufficient basis for the morality of informed consent may turn the doctor–patient relationship into a sort of expert–client offer-and-acceptance transaction. Such an apparently objective type of relationship can imperil the patient's well-being, and even life[4]. Hence, the possible moral risks of the unconditional respect for the principle of autonomy need careful consideration.

One way to counter this objection is to declare that, in practice, the principle of autonomy should be backed up by the principle of beneficence, which helps a doctor supersede the patient's will and choose a certain clinical intervention when the patient's well-being and life are at stake. More precisely, the principle of autonomy should depend on the quality of the patient's choice and its degree of corroboration with the doctor's decision.

Though armed with relevant data provided by the doctor, a patient may decide against a clinical intervention that could improve his/her health and well-being, on the strength of the principle of autonomy. But the doctor could feel justified to take action and protect the patient's well-being, based on the principle of beneficence. In such a situation, the morality of the doctor's intervention would rely on the unconditional respect for the patient's well-being.

Moreover, to increase the legitimacy of the principle of beneficence, it could be argued that, in a therapeutic context, some of the individual's rights should be seen as being limited by his/her status as a patient, those rights being taken over by the doctor in his/her capacity as an expert. In other words, once a patient, the person tacitly admits to their inability to secure his/her well-being and life and cedes the right over his/her well-being and life to the physician. Leaving aside the idealism of

4 We say "apparently" because any expert–client transaction or relationship involves the two sides' subjective experiences and, more often than not, it is subjectivity that prompts their decisions.

the tacitly assumed premise regarding the doctor's unconditional morality, such a strategy will not help us to avoid the objection mentioned above but only lead us to reformulate or transfer it to the other member of the doctor–patient relationship. If unconditional respect for the patient's autonomy could threaten the patient's well-being and life and, therefore, could trigger immoral consequences, then the same could be said about the unconditional respect for the patient's well-being, namely, the principle of beneficence.

Let us, for the sake of argument, suppose that the physician is an ideal moral agent who always bears in mind the Hippocratic Oath and all his/her actions are aimed at the patient's well-being. Even so, just like any human being, the doctor is a fallible cognitive agent who may be wrong in assessing a patient's situation and, therefore, prescribe a wrong treatment. It means that, despite his/her intention to do well, a doctor may put the patient at risk through his/her decisions. Moreover, if, *ex hypothesi*, the result is the only thing that matters to a patient, then the fact that this would be an unintentional error would not change the situation whatsoever. That is because it could be argued equally well that the patient's intentions when he/she rejects a certain intervention recommended by the doctor are just as good as those of the doctor. Moreover, it would be at least difficult, if not impossible, to legitimize the decision of the patient to give up his/her autonomy and free choice in a society based on inalienable individual rights and freedoms. In fact, such an attempt to justify the loss of a fundamental right for the alleged attainment of a greater good will most likely lead to totalitarian regimes.

Another strategy to counter the objection that unconditional respect for the autonomy principle may endanger the patient's well-being is to reinterpret or redefine the doctor–patient relationship. Mazur pleads for the Catholic Church's preeminence in approaching informed consent. Unlike secular doctrines, Mazur says, the Catholic Church promotes a partner-like relationship between doctors and patients. To this effect, he produces a series of Papal documents evincing a generous outlook on the informed consent protocol. For instance, the *Charter for Healthcare Workers* (Pontifical Council for Pastoral Assistance to Health Care Workers, *Charter for Health Care Workers*, Section II, Paragraphs 72–4) describes the relation between doctors and patients as one between equal partners whose subjective experience plays an important role, far better than an expert–client type of relation in which objectivity takes precedence. In other words, the Church sees doctors and patients teaming up to seek the patients' well-being. The same idea pertains to *Ethical and Religious Directives for Catholic Health Care Services*, where the doctor–patient relationship appears as "an important part of the foundation on which diagnosis and care are provided" (*Ethical and Religious Directives for Catholic Health Care Services*, 18), with an emphasis on the personal character of this relationship: "the Church's moral teaching on health care nurtures a truly interpersonal professional-patient relationship" (*Ethical and Religious Directives*, 19). Moreover, as equal partners, doctors and patients have equal, though distinct, responsibilities to preserve a person's well-being: "the person in need of health care depends on the skill of the health care provider to assist in preserving life and promoting health of body, mind, and spirit. The patient, in turn, has a responsibility to use these physical and mental resources in the service of moral and spiritual goals to the best of his or her ability" (*Ethical and Religious Directives*, 18–9).

According to Mazur, this personal type of relationship proves that the process of obtaining informed consent is a process of true personal development. Its main goal is not so much to adequately inform a patient and win his/her consent for a surgical intervention, as to build his/her moral conscience. This is actually Mazur's focal point in favoring the Catholic Church's informed consent doctrine. Though he fails to expand on the conscience formation process, Mazur points out that this element is entirely absent from the most important and significant secular documents on human rights and informed consent. Those declarations make no reference to or acknowledge "the need for formation of conscience" (Mazur 2012, 54).

Ambiguous as this doctrine of conscience formation may be, it certainly does not deal with moral conscience *per se*, but rather with its development in the specific situation of the doctor–patient relationship. In other words, the dialogue between the two partners over informed consent helps both of them to revise their equal moral duties toward a person in need of medical treatment, a process that may lead to a choice in agreement with their duties. Mazur seems to suggest that this equal responsibility toward a moral rule concerning both participants in the dialogue will naturally lead to a positive or, at least, a morally sound, decision. More precisely, this process could help us to avoid the undesirable effects potentially caused by unconditional respect for the principle of autonomy and, correspondingly, for the principle of beneficence. Keeping in mind the doctor's and patient's equally assumed task of preserving the life of any human being, we could say that the two partners are passing the responsibility for the decision and, implicitly, for the result of the clinical intervention, over to the moral rule under discussion. In other words, unconditional respect for the life of any human being turns both the doctor and the patient from decision/action agents into decision/action mediators.

The same interpretation of the relationship between doctor and patient as a partnership and a dialogical relationship determined by their common obligation to respect and protect the life of any human being can be found in the works of other authors, with different accents. Paul Ramsey, for instance, from a more robust Christian perspective, describes informed consent as a partnership based on the joint duty to preserve and protect the life of any human being. He considers informed consent as a joint doctor–patient experience (Ramsey 2002, 11), engaging both in a process of unraveling or, better said, enforcing the Ten Commandments in a medical environment. The doctor's and patient's common duty of care, seen as a sacred virtue sanctioned by the Bible, takes a specific hue in the case of informed consent, as an unconditional respect for the patient's well-being and life.

Unlike Mazur, Ramsey says that the Bible, and not the Catholic Church, is the proper and absolute foundation for our duty to respect the life of any human being. However, the outcome is the same in both cases, because both the doctor and the patient must follow and obey a moral commandment lying beyond the principle of autonomy and the principle of beneficence (which require separate, individual moral duties). Together, they carry out or mediate the divine command.

Another point of view that favors a shared moral foundation that would bind doctor and patient together in the process of obtaining informed consent can be found in Edmund Pellegrino and David Thomasma's work. From a more-nuanced philosophical perspective, the two authors argue that medical practice in general – and informed consent in particular – should be grounded in some fundamental moral virtues that could enforce the morality of both doctor and patient. For example, in *The Virtues in Medical Practice* (Pellegrino and Thomasma 1993, 84), the authors argue that prudence or temperance is a fundamental medical virtue because it helps us take the best ethical decisions applied to each particular situation we come across, knowing that human well-being is the highest aim of any actions. Best ethical decisions are, naturally, those that may, by far, contribute to the patient's well-being. As long as both doctors and patients are educated in the spirit of these basic virtues or observe them unconditionally, the informed consent process is ethically safe.

In other words, similar to Mazur and Ramsey, Pellegrino and Thomasma are looking for a foundation much deeper than the autonomy principle, a foundation capable of making doctors and patients to respect and preserve life as equals. Such a foundation would make informed consent a truly and universally ethical endeavor.

12.4 A moral dilemma and its philosophical presuppositions

The two above-mentioned approaches seem to exclude each other because they build the informed consent process on different foundations and propose completely opposing guidelines. The first approach places the principle of autonomy at the core of the informed consent process, the patient being the sole decision–maker, while the doctor merely facilitates the patient's access to relevant data. In this perspective, informed consent is a mere medical application of the respect for individual liberty. Only the individual can truly decide what is good for him/her. Being a basic human right, it should be observed and respected even when a person's well-being depends on others, as when a person needs medical treatment. To put it in a nutshell, informed consent is seen as a particular expression of individual liberty and freedom of choice.

The other approach claims that the principle of autonomy is not sufficient as a moral foundation for informed consent, and that the principle of beneficence would actually offer a much more solid foundation for the doctor–patient relationship. Still, this perspective does not go to the full length of explicitly denying respect for the patient's autonomy and free choice[5]. However, by stressing that a patient should respect his/her own well-being unconditionally, it implicitly lessens the importance played by the autonomy principle. Therefore, an asymmetrical doctor–patient relationship, which according to the autonomy principle is an objective expert–client tie-up, is replaced by a symmetrical relationship, in which the two equal participants in the decision-making process are bound together by the ethical principle of unconditional respect for the well-being of the patient and, accordingly, for his/her life[6].

At first glance, the antagonism seems to stem from an *a priori* preference for different moral values (freedom of choice versus well-being). If applied, this preference would give birth to divergent rules and procedures. In the case of the first perspective, unconditional respect for human well-being would depend on individual autonomy. For the second perspective, things would work the other way round, with respect for human well-being taking precedence over respect for the patient's autonomy. But such a description would be too straightforward and leave behind a natural and highly important question: why the preference for distinct values? All the more so as

5 That could be easily obtained by enforcing the beneficence principle as necessary and sufficient for medical practice. As a result, doctors would always and unconditionally have to act in favor of the patients' well-being, regardless of the latter's individual choices.

6 Someone might argue that the unconditional respect for human life is very different from the unconditional respect for the well-being of a patient and, accordingly, that they should not be taken together here. The first kind of respect is based on the principle that asserts, from a religious perspective, the sacredness of life (or, from a secular point of view, that life is an end in itself), whereas the second one is based on the principle of beneficence. But it is very difficult to see how a doctor could act according to the principle of beneficence (which means to act so as to always promote and sustain the patient's well-being) without respecting, at the same time (even if implicitly), the life of the patient as an end in itself or as being sacred. One usual reply to this is that, in the case of assisted suicide, the action of a doctor who puts an end to the patient's life (and, therefore, to his or her suffering) could be seen as morally legitimate according to the principle of beneficence, whereas, if the doctor would respect unconditionally the principle that asserts that life should be valued no matter what, he would not do this action. But how could this action be legitimate as an action that intends the well-being of a person? Well, the usual answer runs like this: by putting an end to suffering. The fact is, though, that after such an action, there would not be any patient/person who could benefit from that action. Assisted suicide is somehow a maximal or limit-case procedure, and it cannot be judged as we judge a surgical intervention to remove a damaged limb, for instance. In the second case, the patient will continue to exist and will be able to benefit from that action, whereas in the first case, this is obviously not true. So, we could say that the principle of beneficence could not be used, properly, to justify such an action, but rather the principle of autonomy. More precisely, the action of a doctor who acts according to the patient's will to suspend his or her life is legitimate in that it respects the right of the patient to decide over his or her own life, even if it may not be legitimate according to the principle of beneficence. If we accept this line of reasoning, we could say that it is rather natural and legitimate to associate the respect for the well-being of a patient with the respect for human life, and not vice versa.

this preference manifests itself both at the level of foundations and also that of the rules and procedures guiding the process of obtaining informed consent.

For example, in the first perspective, a doctor may resort to arguments and appeal to the patient's reason in order to persuade him/her to accept a certain medical intervention. Unlike coercion, persuasion does not erode the respect for autonomy, because "persuasion is restricted to influence by appeal to reasons" (Faden and Beauchamp 1986, 261), which means that a medical doctor may *plead* for a certain intervention by presupposing that the patient is a reasonable-enough person to choose the best solution based on solid reasons.

In the second perspective, based on respect for the patient's well-being and life, the appeal to reason is replaced by an appeal to a common set of values, which should be observed unconditionally by the physician and his/her patient. This approach appears quite discernible in Mazur's analysis. While describing conscience formation as the main goal of the doctor–patient dialogue, Mazur takes a harsh stand against both human rights documents and secular theories on informed consent, criticizing their ignorance of ethical counseling. For instance, referring to human rights documents on informed consent, he notices that "none of these documents ever used the word 'conscience' except for the Declaration of Helsinki (...). Still, the Declaration never refers to the conscience of the patient, which is a substantial shortcoming" (Mazur 2012, 54). Later on, when discussing Beauchamp and Childress's theory, he concludes by saying that "although our authors allow for counselling in the informed-consent process, they understand it narrowly and do not provide satisfactorily for conscience or for its formation." (Mazur 2012, 71).

In the context of these considerations, we can introduce the following explanatory hypothesis: the answer to the above-mentioned question depends on the philosophical presuppositions that support the two perspectives on informed consent. The penchant for radically different values that characterize the two viewpoints is not a matter of free choice or, better said, it is not a preference, as we called it. It is rather a natural consequence that follows tacitly from two opposing worldviews and, in particular, from two opposing views on human nature.

In the first case, respecting the patient autonomy and founding the informed consent doctrine on the principle of autonomy emerge, naturally, from a secular vision of human nature. Individual freedom, including a person's freedom of choosing what to believe about himself/herself and his/her nature, is an innate, inalienable right. Individuals choose what is in their best interest to preserve their well-being and, in the specific framework of medical care, whether a medical intervention is acceptable. In short, only individuals have the right and freedom to decide what to do with their lives.

Correspondingly, in the second case, the respect for the patient's well-being and the justification of informed consent on the basis offered by the principle of beneficence stems, more often than not[7], from a deeply religious worldview. Life is a gift given to man by God and unconditional respect for life is, therefore, the basic duty of every individual to God. That is why, in the context of the doctor–patient relationship and of the informed consent process, a God-fearing patient is bound to put his/her duty to respect life *before* his/her fundamental right to the freedom of choice. Moreover, given the fact that the patient's religious duty is also shared by his/her doctor, this gives rise to what could be called a symmetry of participation, leading to a subjective cooperation between doctor and patient or, as Mazur called it, the formation of moral conscience.

This hypothesis, according to which the values and, implicitly, the distinct principles on which informed consent is founded originate in different philosophical views, has at least two important consequences, one theoretical and the other, rather practical.

The first consequence is that the two perspectives on informed consent are irreconcilable and, perhaps more importantly, incommensurable. Since they stem from opposing philosophical presuppositions, any attempt at intertwining them or, as we have seen, criticizing them or pointing out their weaknesses against each other, is doomed to failure. Once the philosophical presuppositions behind them are made explicit, such debates become superfluous or simple rhetorical exercises. The reason can be easily detected: against the background offered by a secular metaphysical worldview, one according to which freedom is the best individual asset, the autonomy of conscience is a nonnegotiable, if not absolute, value; and vice versa, against the metaphysical background offered by a religious worldview, according to which human beings are essentially divine, the respect for human life becomes a decisive, nonnegotiable command.

In the first case, the necessity of (moral) conscience formation through the dialogue between the practitioner and the patient seems to be unjustified, if not extravagant. And that is because this secular worldview sees a person and his/her defining traits, including the formation of moral conscience, as a private matter, strictly reserved for individual autonomous introspection.

In the second case, if we assume the background offered by a religious worldview, the lack of such a process of moral counseling in the protocol of obtaining informed consent could seem, indeed, revolting[8]. That process could help the doctor and the patient to become aware of their moral duties or to revise them in the context of the divine command.

7 Although not necessarily, of course. Life could be seen as an end in itself on purely secular grounds, but the resulting effect would be the same, both the doctor and the patient would be bound to act so as to preserve and maintain life unconditionally.

8 As Mazur sometimes seems to be, alarmingly and repeatedly, pointing to this omission in the secular theories on informed consent, despite their own merits.

The second consequence refers to medical practice, more exactly to the ethical uncertainties and traps that may affect the process of obtaining informed consent if we ignore the two antagonistic theories. As we have already seen, the two doctrines require an honest and truthful doctor–patient dialogue to help provide adequate information and lead to the patient's consent for the proposed medical treatment or clinical intervention. At first sight, this common trait could be neutral or constant, regardless of the above-mentioned theories. Informed consent will always involve communication, dialogue between doctors and patients. The content of that dialogue will always focus on a specific context the doctor and patient are in. But a deeper analysis will reveal that the dialogue and the direction where it heads depend, in their turn, on the metaphysical presuppositions of the informed consent doctrine assumed by the participants in that dialogue. To be more specific, a doctor may commit himself/herself to a process of rational persuasion or moral counseling of the patient, according to his/her own theory about informed consent. The patient may naturally react according to his/her own vision on this subject. We could easily imagine the confusion and disagreements that might appear and which could seriously undermine the doctor–patient relationship, be it a supposedly objective expert–client type relationship or a subjective fair-and-square equal partnership.

A dialogue that refuses to acknowledge and make explicit both sides' metaphysical presuppositions may never reach the consent stage or it may lead to imposing a particular moral ideology on the patient, disguised as a universal ethics. Both situations would eventually cripple the ethics of the whole process of informed consent and, sometimes, the patient's autonomy or well-being.

12.5 Conclusions

The solution to this theoretical and practical deadlock follows, somehow naturally, precisely from the moral dilemma mentioned above, and it already lies in the two doctrines discussed. If the ethics of the informed consent process depends on the worldviews of the two participants to the dialogue, then they should first of all go through a philosophical dialogue or counseling process to explain to each other and make explicit their personal metaphysical worldviews. This could take place in the presence of a counselor with philosophical expertise, who could act as an arbiter and could help the doctor and the patient to identify and properly formulate their particular views and the arguments on which they rest. Once this process of philosophical counseling is accomplished, the doctor and the patient may choose, together, the most suitable type of decision-making procedure in seeking genuine informed consent. At the end of this, nobody will be able to deny that consent was sought on a proper ethical background.

At first sight, our proposal may seem excessive, even extravagant. Why would physicians and researchers add one more stage to the process of seeking informed consent, thus delaying the patient's choice and the medical intervention? Although we cannot afford to reply to this objection at length here, it is worth noting that our proposal rests, implicitly, on the assumption that such a philosophical dialogue should be carried out only when time is not critical, and the life of the patient is not in imminent danger. In cases where there is simply not enough time, the discussion would be, of course, superfluous and inefficient. In such cases, even the protocol of obtaining informed consent would be disregarded, and the doctor would take a decision based solely on his/her medical expertise. But in cases where time is not a pressing factor, this preliminary philosophical dialogue between the doctor and the patient could play a critical role, considering the crucial importance of seeking informed consent. No measure can be disregarded as excessive or extravagant if we have time to use it to bring significant benefits or save a person's life (including here the person's right to have his/her own philosophical beliefs about human nature and the world). To uphold an opposed viewpoint would implicitly mean to disregard what is at stake, to shun the importance of seeking informed consent. Indeed, it would be difficult to sustain such a viewpoint on ethical grounds, and, accordingly, it would be unacceptable, particularly in the context of the subject tackled by this study.

References

Ethical and Religious Directives for Catholic Health Care Services, United States Conference of Catholic Bishops, 18. http://www.usccb.org/issues-and-action/human-life-and-dignity/health-care/upload/Ethical-Religious-Directives-Catholic-Health-Care-Services-fifth-edition-2009.pdf

Faden, Ruth F., and Tom L. Beauchamp. 1986. *A History and Theory of Informed Consent.* Oxford: Oxford University Press.

Katz, Jay. 1972. *Experimentation with human beings. The Authority of the Investigator, Subject, Professions, and State in the Human Experimentation Process.* New York: Russell Sage Foundation.

Katz, Jay. 2002. *The Silent World of Doctor and Patient.* Baltimore: John Hopkins University Press.

Mazur, Grzegorz. 2012. *Informed Consent, Proxy Consent, and Catholic Bioethics.* New York: Springer Science+Business Media B.V.

Pellegrino, Edmund, and David C. Thomasma. 1993. *The Virtues in Medical Practice.* Oxford: Oxford University Press.

Pernick, Martin S. 1982. "The Patient's Role in Medical Decision making: A Social History of Informed Consent in Medical Therapy", President's Commission for the Study of Ethical Problems in Medicine and Biomedical and Behavioral Research, *Making Health Care Decisions.* Washington, D.C., U.S.: Government Printing Office.

Pontifical Council for Pastoral Assistance to Health Care Workers, *Charter for Health Care Workers*, section II, paragraphs 72-74.http://www.ewtn.com/library/curia/pcpaheal.htm

Ramsey, Paul. 2002. *The Patient as Person: Explorations in Medical Ethics.* New Haven: Yale University Press.

Cristian Iftode

13 Bioethics as Biopolitics: A Foucauldian Perspective

13.1 Introduction

Talk of bioethics as biopolitics usually relates to the common idea that debates about the quality of life, medical ethics, or the advances in biology and neuroscience are influenced by a variety of preexisting political positions (Reiner 2013). Political views about how society should be organized have significant implications for practical issues related to reproductive freedom or public health. For instance, liberal positions tend to favor the moral permissibility of human enhancement as long as it does not violate individual freedoms (Agar 2004), whereas bioconservatives are more skeptical (Sandel 2007).

In this article, I draw attention to a different interpretation of bioethics as biopolitics, which does not appeal to the standard application of political theories to controversial practical issues. My objective is to make several suggestions for approaching bioethics from a Foucauldian perspective. I follow the three stages in Foucault's intellectual trajectory (with a focus on the genealogy of power) and also analyze the way he reflected upon themes that are part of bioethical issues. The French philosopher is famous for his relentless investigations to uncover insidious forms of power and how scientific discourses can be used to reinforce or restructure social orders[1], highlighting the dangers of scientism while claiming to overcome the "critique of ideology" approach. Such an analysis is especially needed, as Foucault's ideas are often linked to the field of bioethics in a rather distorted, incomplete, or fragmentary way[2], often as secondhand quotations, lacking a thorough and careful

1 What brings a particular scientific discourse "in addition to the real", instead of merely "representing" reality? (Foucault 2017, 235–9) Focusing on "practices constitutive of domains of objects and concepts" (Foucault 2014, 12), his approach of the human sciences was, broadly speaking, a pragmatist one.

2 See, for instance, Bishop and Jotterand (2006), who, mistakenly, argue that Foucault would have maintained a somehow positive – or at least "ambiguous" – view of "the enabling capacity of biopolitics", without acknowledging the common dark core of modern democratic societies and totalitarianisms revealed by Agamben (1995). In fact, a text such as "The Subject and Power" (Foucault 1983) clearly highlights "the strange, perverse, insidious alliance between the effect of *totalization* and the effect of *individualization* that lies at the very heart of Western democratic societies" (Iftode 2012, 96–7). Fascism and Stalinism, the two great "diseases of power" of the 20th century, would have actually "used and extended mechanisms already present in most other societies. More than that: in spite of their own internal madness, they used to a large extent the ideas and the devices of our political rationality" (Foucault 1983, 209).

study of his understanding of power techniques, the human subject, personal freedom and society, medicine, and sciences of life.

13.2 Madness, the asylum institution, and the psychiatric power

Foucault's first major work, *Histoire de la folie à l'âge classique* (1961), provided a negative answer to a question that became central to the antipsychiatry movement developed in the 1960s: is madness (*folie*) really reducible to mental illness?

The French thinker argued that, in early modern times, we witnessed the suppression of the ancient and medieval complementarity between reason and madness (invoking Apostle Paul's warning, any reason holds its degree of madness and any madness holds its reason – 1 Cor 1:18-25). He also claimed that this suppression parallels, on the level of ideas, the exclusion of mad people from the social sphere, starting with the Great Confinement of 1656 (Paris, L'Hôpital Général). The social exclusion would have gone hand in hand with the "Cartesian" decision of denying the mad person any kind of moral or intellectual status. Based on "methodic doubt", someone can suppose that s/he is now dreaming, or that the world is an illusion, but s/he cannot doubt the fact that s/he is not mad, because "madness is precisely a condition of impossibility for thought" (Foucault 2006a, 45). This assumption involves *silencing* the mad, breaking any possible connection between sane reason and unreason, the latter being understood simply as "madness".

But what is more important to us, on a social and moral scale, is the creation of hospital institutions as "a sort of semijudicial structure", "an instance of order", organized on the basis of "former lazar houses" in major cities in France, during the 17th and 18th centuries. What becomes essential in Western societies (either Catholic or Protestant) is this obligation to work: "Once, he was welcomed because he came from without; now he was excluded because he came from within, and the mad were forced to take their place alongside paupers, beggars and vagabonds" (Foucault 2006a, 62). What is now important is eliminating any factor that threatens social order: "Confinement (...) was a 'police' matter (...) quite independent of any desire to cure. What really made it necessary was a work imperative" (Foucault 2006a, 62). And this is how the idea of *discipline* as the right way of "reclaiming irregular men" (Foucault 2006a, 76) comes into play.

Taking into consideration the next stage of this "history of madness" (the end of the 18th and the 19th centuries), what is there to say about the "liberation of the mad" in chains, accomplished by "philanthropists" such as Tuke in England or Pinel in France? Foucault understood this historical episode as the symbolic expression of the modern, scientific way of reestablishing the bridge between reason and madness, but only by assuming reason as the "truth" of madness and conceiving the psychiatric treatment as negating ("sublating") the (mental) alienation. This means that the whole purpose of the psychiatric treatment consists in getting the mad person to speak and behave again in a civilized, "normal", and socially acceptable manner.

But while trying to "tame" the madness, to make it listen to the voice of reason, the Enlightenment would actually prove its dark side: the other side of madness is the madness of *discipline*, the madness of the guardians, of the keepers, relishing the "beast" in the very name of order and discipline: "animality was not to be found in the animal, but in its taming" (Foucault 2006a, 477).

Still, the reasons for doing that were of the noblest kind. It is clear, at this stage, that the psychiatric cure was in fact a *moral* treatment, working on the madman's guilt and carefully organizing it, aiming at "ethical uniformity" through "this conversion of medicine into justice, and therapeutics into repression" (Foucault 2006a, 493, 501). Consequently, Foucault emphasized a number of *disciplinary techniques* and *strategies of normalization*[3], which were specific to lunatic asylums: silence, recognition as mirror, perpetual judgment, and most importantly for the future of psychotherapy, the apotheosis of the medical character (*personage médical*). He noted, "The doctor could only exert his absolute authority over the world of the asylum in so far as he was, from the beginning, Father and Judge, Family and Law" (Foucault 2006a, 506).

The shape of psychiatric practice pertaining to social morality and the positivist foundation of modern psychiatry, strangely combined with a mythical aura of the psychiatric physician, make psychiatry unable "to hear the voices of unreason"[4], concluded Foucault.

Intended as a laboratory for a second volume of *History of Madness* that was never to be accomplished[5], *Psychiatric Power: Lectures at the Collège de France 1973–1974* reflects Foucault's changed focus on the genealogy of power relations. Instead of starting from an analysis of cultural representations about madness or the "perception of madness", Foucault was concerned right from the start with "the apparatus (*dispositif*) of power as a productive instance of discursive practice" (Foucault 2006b, 12–3). Regarding the psychiatric hospital from a Nietzschean perspective, a conceptual reshuffling is now required: rather than speaking of "violence", he speaks of "a microphysics of power"; instead of "institutions", he tackles the "tactics" of the psychiatric power; and instead of comments on the "family model" or the "State

---

3 I am using two expressions that do not actually appear in this book but will be made popular by Foucault later in the 1970s.

4 "Madness need not be all breakdown. It may also be breakthrough (...) The person going through ego-loss or transcendental experiences may or may not become in different ways confused. Then he might legitimately be regarded as mad. But to be mad is not necessarily to be ill, notwithstanding that in our culture the two categories have become confused" (Laing 1971, 133, 138).

5 The final part of the lectures somehow merged into the elaboration of the first volume of *History of Sexuality* (1976/1978), while the analysis of power *dispositifs* was undertaken in *Discipline and Punish* (1975/1995).

apparatus", he is now discussing "the strategy of these relations of power and confrontation which unfold within psychiatric practice" (Foucault 2006b, 16).

Foucault's investigation amounted to a most disturbing paradox regarding the functioning of a modern psychiatric hospital: unlike any hospital of general medicine, "the psychiatric hospital exists so that madness becomes real" (Foucault 2006b, 252). But his analysis goes one step further than the institutional type of criticism of the psychiatric hospital, revealing a peculiar *double bind*: the purpose of the asylum institution remains a suppression of "the symptoms of madness", but, at the same time, the psychiatric power regards the asylum as the "space of realization" for the mental illness, in order to justify the restraint of the patients (Foucault 2006b, 252–3).

In Foucault's view, the psychiatric power was not essentially repressive: it was a way of *producing* a particular kind of knowledge, the scientific classifications and discourses about mental disorders justifying the fundamental *division* of speeches and conducts into "normal" and "abnormal" ones. Thus, we should be aware of the potential of psychiatric knowledge to shape social institutions, which in turn "normalize" collective behavior.

13.3 Biopolitics as "power's hold over life": the three levels of an analysis of "normalization"

In June 2016, the UN World Health Organization (WHO) issued a public warning to Syrian officials, demanding their collaboration in order "to control the use of tobacco and water pipes" among Syrian people. Although according to Dr. Elizabeth Hoff, WHO's Syria representative, the use of water pipes to smoke shisha "is 20 times more dangerous than cigarette smoking", controlling the use of cigarettes remains WHO's main target, "presumably because they can't apply their risible plain packaging policy to water pipes" (Snowdon 2016). According to an Associated Press report, "Syria's war is estimated to have killed several hundred thousand people amid the rise of the Islamic State group. But Dr. Ahmad Khlefawy, Syria's Deputy Minister of Health, said the war cannot be an excuse for Syrians to endanger their lives by consuming tobacco." (Associated Press 2016)

To complete the dark irony of the situation, it has to be said that the most effective tobacco control strategy is already in place in the territories controlled by the Islamic State, where smoking is banned, with punishments ranging from whipping to execution. This being the case, the act of smoking, however bad it is for health, comes to be seen by Syrian people as a kind of symbol for personal freedom and rejection of Islamic fundamentalism.

This could be a perfect example of what Foucault envisaged when using the term "biopolitics". But what does *biopolitics* actually mean? The term was probably coined in the 1920s by the Swedish political theorist Rudolph Kjellén and initially used by authors who were hostile to the liberal contractualist theory to designate a conception of the State as a supraindividual "form of life" (Lemke 2011; Esposito 2008). Later used by Nazi ideologues (such as Hans Reiter) with explicit racist connotations (Liesen and Walsh 2011), the term "biopolitics" was reintroduced in social sciences at the beginning of the 1960s, in the context of discussions of the various aspects of political behavior in relation to psychobiological and neurophysiological research. Therefore, it has to be said that by the time Foucault made it notorious, the word "biopolitics" had already been used in many contexts.

Thus, it is important to understand how exactly Foucault *reshaped* the meaning of "biopolitics". His famous definition appears in the last chapter of *The History of Sexuality I,* which was initially published in 1976:

> [We] speak of *bio-politics* to designate what brought life and its mechanisms into the realm of explicit calculations and made knowledge-power an agent of transformation of human life (...) For millennia, man remained what he was for Aristotle: a living animal with the additional capacity for a political existence; modern man is an animal whose politics places his existence as a living being in question. (Foucault 1978, 143)

Already at use in a series of conferences conducted in Rio de Janeiro, in 1974, about the origins of "social medicine" (Foucault 2001a, 210), "biopolitics" would become a key notion in the lectures delivered at the *Collège de France* between 1976 and 1979 (Foucault 2003b, 2008, 2009). Instead of analyzing to what extent political institutions are effects of our biological condition, the French thinker questioned whether modern times are not synonymous with an era when the bare *life* of human beings becomes the primary target of *politics*, thus enabling a whole range of strategies, techniques, and mechanisms for the "management" of life and the enhancement of our natural traits.

These *control* strategies can operate directly and openly as it happened in the case of 20th-century catastrophic totalitarianisms. But they can also function in a subtler and implicit manner, in the context of Western liberal democracies, as indirect ways to generate various regulations, fiscal policies, funding policies for research deemed to be useful in terms of general social objectives, and strategies for initiating public debates on legislative changes, or more generally, for inscribing particular topics on the public agenda. To put it briefly, let us suppose that the "neutrality" assumed by the state power will allow only the smallest possible number of direct prohibitions and consider violent restraint merely as an extreme solution. This fortunate situation would still imply that state power does exercise itself through tactics aimed at continuously creating a particular kind of "demand", a public anxiety and expectation. This would equally imply that "politics" is actually guiding the individual pursuit for happiness by emphasizing and valuing a specific type of success, of achievement, of conduct to be followed.

During the 1970s, pursuing a Nietzschean thread, Foucault focused on a genealogy of the modern subject understood not as the grounding principle of all human knowledge and action, but as *sujet de pouvoir*: an *effect* of power relations that fold the self, force "the individual back on himself and ties him to his own identity in a constraining way" (Foucault 1983, 212). The general strategy was to show how both "the subject of law and legal theories and the *normal* man of human sciences" are being produced through subjection (*assujettissement*) by a "disciplinary power that informs our practices, induces docility and regularity, normalizes conducts" (Gros 2004, 19). So, in the first stage, "normalization techniques are understood only in the manner of disciplines" (Le Blanc 2006, 155). Foucault talks about a "microphysics of power" that functions in various institutional frameworks: the asylum (as we have already seen) or the prison, but also factories, military barracks, school classrooms, or public administration – each of these cases involving precise disciplinary techniques (Foucault 1995). But starting from 1976, he focused on the question of "bio-power" or "biopolitics" conceived as "power's hold over life", "the acquisition of power over man insofar as man is a living being", or even "State control of the biological" (Foucault 2003b, 239–40). This was Foucault's way of addressing what many Leftist intellectuals of that time considered was previously missing from his approach: a genealogy of modern state power and state apparatus, both different from traditional sovereign power[6] and irreducible to a disciplinary power operating on the microlevel of particular institutions such as the ones mentioned earlier.

Even if his reflections lack articulation of an actual "theory" of political power, they gradually merged into a vision of modern power relationships forming a complex mechanism that would have developed in the Western world during the 17th and 18th centuries, which involved two dimensions. First, this "multiple, automatic and anonymous power" (Foucault 1995, 176) over life focused on the *disciplines* of the body: a whole range of techniques designed to train (in fact, "tame") working individuals who are useful and docile. Then, toward the end of the 18th century, we witness "a second seizure of power that is not individualizing but (...) massifying, that is directed not at man-as-body but at man-as-species" (Foucault 2003b, 243). In the latter case, the technology of power does not involve direct threats or physical corrections but resorts to *regulations* of the population – a population that is conceived, at the same time, as a social body and "species body (...) imbued with the mechanics of life and serving as the basis of the biological processes" (Foucault 1978, 139). *Biopolitics* establishes itself through specific devices of power/knowledge, such as the newly emerged *statistics*, the "science of the State", and it involves new practices in "health management, hygiene, nutrition, sexuality, birth rate, up to the point where these have become political stakes" (Revel 2002, 13). This correlation of disciplines and regulations (as two mutually implying technologies of power) gives us the key to grasping the troubling thesis formulated by Foucault later in the 1980s:

6 "One might say that the ancient right to *take* life or *let* live was replaced by a power to *foster* life or *disallow* it to the point of death" (Foucault 1978, 138).

> [T]he state's power (...) is both an individualizing and a totalizing form of power. Never, I think, in the history of human societies – even in the old Chinese society – has there been such a tricky combination in the same political structures of individualization techniques, and of totalization procedures. (Foucault 1983, 213)

In this context, the key concept and also the most disquieting one seems to be the idea of "biopolitical normalization". What is the meaning of "norms" and "normalization" according to Foucault? How are we to understand his proffered "antinormativism"? The issue is of the highest complexity, but I would say that the most important thing, at least from a methodological point of view, is Foucault's dismissal of the *ideality* of norms (Legrand 2007, 153–5). Once this position is assumed, his strong rejection of a disciplinary society, as well as his criticisms of the gradual "juridification" or "very strong 'codification' of the moral experience" (Foucault 1990, 30) in the Western world, follows as a matter of consequence. Norms are never to be conceived as expressions of "Divine Will" or "Pure Reason", "but rather as material statements (*énoncés*) acquiring a normative significance within precise frameworks of action and through distinct social practices" (Iftode 2015, 146).

The *biopolitical* age is synonymous with this historical overlapping of the *juridical*, coercive meaning of the norm, the *biological* meaning (the "normal" functioning of an organism), and the *statistical* meaning ("normality" as an average). This is already quite obvious in the lectures about the *Abnormal*: "the norms function is not to exclude and reject. Rather, it is always linked to a positive technique of intervention and transformation, to a sort of normative project" (Foucault 2003a, 50). So, in a genealogy of abnormality, we shall have, on the one hand, the "individual to be corrected" and, on the other hand, the "monster" seen as an anomaly, the "natural form of the unnatural". In this context, "the recurring problem of the nineteenth century is that of discovering the core of monstrosity hidden behind little abnormalities, deviances, and irregularities" (Foucault 2003a, 56).[7]

However, the approach later undertaken in the lectures on *Security, Territory, Population* (Foucault 2009) allows us to distinguish *three* levels of "biopolitical normalization": *law*, *discipline*, and *security*. To be more precise, we have to distinguish *the legal system* (involving "a binary division between the permitted and the prohibited", and also a link between prohibition and punishment), *the disciplinary mechanism* (involving techniques of surveillance, correction, and so on), and *the apparatus (dispositif) of security*. As to the functions of an apparatus of security, we might again distinguish three of those: (a) to insert a phenomenon "within a series of probable events"; (b) to insert "the reactions of power to this phenomenon" in "a calculation of cost"; (c) to establish, "instead of a binary division between the permitted and the prohibited", something like "an average considered as

7 See Lombroso (2006), the classical study of the "born" criminal, originally published in 1876.

optimal on the one hand, and, on the other, a bandwidth of the acceptable that must not be exceeded" (Foucault 2009, 20–1).

Foucault explains these three different levels of normalization by giving two kinds of historical examples. One is the punishment for theft, while the other directly targets the field of bioethics: the treatment of leprosy, plague, and smallpox. For leprosy, we encounter a *legal* dividing practice aimed at the exclusion of lepers, while in the case of plague, there existed *disciplinary* regulations indicating where and when you can go out, and also prescribing a particular conduct at home, a food diet, the avoidance of some types of personal contact, and the obligation to allow regular inspections in your house. But in the case of smallpox, we may witness, in the 18th century, the emergence of *security* procedures directly linked to "knowing how many people are infected with smallpox, at what age, with what mortality rate, lesions or aftereffects, the risks of inoculation, the probability, and the statistical effects on the population in general" (Foucault 2009, 24).

Foucault does not hold that the mechanisms of security are something new; neither did he claim that they would involve a "cancellation of juridico-legal structures or disciplinary mechanisms". What he actually achieved was asking whether we have begun living in a "society of security" (Foucault 2009, 25), one in which "basically, the fundamental question is economics and the economic relation between the cost of repression and the cost of delinquency" (Foucault 2009, 23). In the lectures dating from the following year (Foucault 2008), he openly addressed questions regarding American and German neoliberalism, where what is at stake is not only a kind of *laisser-faire*, a certain "freedom of movement (*laisser-passer*)" and a sort of "letting things take their course" on the market (Foucault 2009, 64), but also an attempt to understand all our private and public relationships (and first of all the relationship to the self) on the ground of a particular economic model. The full replacement of "*homo œconomicus* as partner of exchange with a *homo œconomicus* as entrepreneur of himself" or as "enterprise-unit" (Foucault 2008, 225–6) leads to the general view toward oneself as holder of a "human capital" (both innate and acquired). Then, the purpose of existence becomes the attempt to fully benefit from this capital.

We are witnessing today extensive debates on how adopting this "neoliberal" pattern of thinking alters all aspects of personal and public life (such as education or love relationships) (Dardot and Laval 2009). But what is even more interesting in this context is Foucault's foresight of what would become a key issue in the present debates around *transhumanism* and human enhancement: addressing the question of the morality and the availability of possible interventions aimed to improve the genetic makeup of individuals basically in terms of the *costs* of these procedures.[8]

A possible reply to Foucault's critical analysis of American neoliberalism could be the following: what Chicago School members are actually doing is making use of economic theory in order to provide a fundamental model for *describing* social behavior, without formulating *normative* claims. However, the key point is understanding that Foucault's idea of biopolitical normalization – conceived as the historical consequence of blending the juridical, the biological, and the statistical meaning of norms – makes highly problematic *the distinction itself between descriptive and normative*. There is no such thing as a *neutral* description of social phenomena. A clear-cut separation between descriptive and normative may exist only if we are holding on to that *ideal* nature of norms that Foucault clearly rejects. For this very reason, once we acknowledge that the description of all social realities through concepts borrowed from economic liberal theory is becoming more and more "natural" in our times, it is quite naïve or politically questionable not to realize that these descriptions will affect the whole range of social relationships, as well as the relationship to the self. We often hear that it is "normal" to behave like this, calculating the costs of any endeavor or interaction and trying to maximize personal gain. We are being told that this is the way a *rational* agent would behave in any particular situation. So it becomes almost impossible not to "bend" to this generic description of a human *individual*. In an age when the *telos* of the existence comes to be seen solely in terms of what is measurable, quantifiable, and reducible to the "horizontal" of physical *health*, economic *welfare*, and social *security*, the *human enhancement*, conceived as a way of increasing the "human capital", becomes nothing else than what increases an individual's chances of social success, thus contributing to the general welfare. We strive to be more attractive, healthier and more resistant, better informed and up to date, easily adaptable, and more cooperative.

Further clarifications may be required, but I shall only mention one more practical distinction on the question of normativity, advanced by Foucault in his 1978 lectures:

> Due to the primacy of the norm in relation to the normal, to the fact that disciplinary normalization goes from the norm to the final division between the normal and the abnormal, I would

8 "The traditional terms of racism" are not suitable anymore "at the level of actuality", where biopolitics has become inextricably linked to capitalist economy. The use of genetics has become a problem of "the formation, growth, accumulation, and improvement of human capital" (Foucault 2008, 228). It is important to notice that a key sentence from Foucault's 1979 course is actually missing from the English translation: "*Et vous voyez très bien comment le mécanisme de la production des individus, la production des enfants, peut retrouver toute une problématique économique et sociale à partir de ce problème de la rareté des bons équipements génétiques*" (Foucault 2004, 234). So, it may come to this: if you want your offspring to have a good start in life, you must have the means to *invest* in their superior genetic makeups.

> rather say that what is involved in disciplinary techniques is a normation (*normation*) rather than normalization. (Foucault 2009, 85)

As to the apparatuses of security, a norm is not something primarily given, but something that is reached through "statistical instruments" and "the calculus of probabilities"[9]: "Thus we get the idea of a 'normal' morbidity or mortality" (Foucault 2009, 90) from, let us say, smallpox. It is about establishing "acceptable limits", rather than "the imposition of a law that says no to them" (Foucault 2009, 93). And it is here we can properly speak about "biopolitical normalization" and, in close connection to this, about "utilitarian philosophy" as "the theoretical instrument" for this new *management* or "government of populations" (Foucault 2009, 102).

From a genealogical perspective, "biopower takes over the activity of care of the self" (McGushin 2007, 238). Conceived as the practical goal of ancient virtue ethics, "self-care" first found itself taken over by the pastoral power during the Middle Ages, in the shape of the "government of souls" and then was later transferred to the new political structure of the modern State, starting with 17th-century Western Europe. Drawing on Foucault's four-fold conception of ethics (I shall come back to this right away), McGushin (2007, 238–9) suggested that we may grasp the structure of biopower or biopolitics using the same framework: (a) the focus is on "the productive biological substance of life", (b) the relationship to the rules is established through what might be called "rational choice" (you follow the rules because you are told this maximizes your individual prospects of survival and wealth), (c) your identity is shaped through disciplines and regulatory controls, and (d) the social goal (*telos*) is normalization.

13.4 In search of a new "ethics of life": ethical subjectivation vs. political subjection

Considering the contents of the previous section, I hope that it was clear that Foucault's so-called "ethical turn" from his final years must be placed in the context of the idea of resistance to "biopolitical normalization":

> [I]t is perhaps an urgent, fundamental, and politically indispensable task, that of constituting an ethic of the self, if it is true that after all there is no other point (…) of resistance to political power than in the relation of the self to itself. (Foucault 2005, 251–2)[10]

---

9 See Foucault's (2009, 90) discussion of the innovative medical practices of variolization and vaccination.

10 The passage is actually quoted from McGushin (2007, XV); Foucault (2005) offers a slightly different translation of it.

We have to be aware of a number of things when reading this declaration. First of all, we have to be aware of the fact that Foucault (1990) actually reshapes the meaning of "ethics" as the practical way of self-formation, i.e., of establishing and maintaining a particular relationship to the self (*rapport à soi*). This is to be contrasted with the already-mentioned strong "juridification" or "codification" of the Western moral experience and of the modern moral philosophy (Foucault 1990, 30), focused on our relations with others. In this context, he distinguishes those four elements of moral self-constitution that McGushin was considering: the ethical substance that has to be shaped; the mode of subjection (*assujettissement*), depending on a particular understanding of the nature of moral rules; the *ethical work* that one performs on oneself; and the *telos* of the ethical subject. Foucault is advancing a compelling version of ethical *pluralism*, where the variety of "arts of living" is ultimately explained by the existence of multiple and divergent understandings of what a moral rule is. His controversial conception of ethics as "an aesthetics of existence" becomes in this way more intelligible: far from being a plea for irresponsible dandyism, it is about "the conscious (*réfléchie*) practice of freedom" (Foucault 1997, 287), but of a freedom that remains, in its primary meaning, something of a prereflexive "instinct" to resist external constraints.

I judge Foucault to have been a strong supporter of "negative" freedom, but one who did not distinguish between "two concepts of liberty" (Berlin 1969), implying instead the existence of two sides or two moments of the same movement that leads from the rejection of discipline to *self-discipline*. It remains true that the primary expression of freedom is an expression of independence, involving the rejection of discipline and even the attempt "to get free of oneself" (Foucault 1990, 8) and reject any sense of identity settled once and for all. However, from an ethical perspective, we witness how this "No" is then forced to convert itself into a personal choice and self-regulation of a particular life discipline, this being the only way of not letting yourself be driven by chaotic and self-contradictory momentary impulses.

The key to this challenging view is provided by Foucault's extremely subtle understanding of the complicated interplay between power and freedom, *subjection* and *subjectivation* in our lives. The unsettled and undecided nature of our identities makes possible, at any time, the resistance to normalization and social conformism, while making uncertain, if not utterly impossible, that *definitive* "printing" of traits required by traditional virtue ethics[11]. Still, what seems important, from an ethical point of view, is to convert the "instinct" of freedom into the *freedom to give yourself rules of conduct*, as in the case of the aesthetic choice through which a work of art

---

11 For a relatively similar interpretation of ethico-aesthetic subjectivation as "an ephemeral, never to be completed work-in-progress", refer O'Leary (2002, 133).

gets done[12]. When you go through Foucault's interpretation of ancient philosophical "instruction" (*paraskeuê*) and *askêsis*, you may have the impression that his final lesson is that the moment of choice is nothing more and nothing less than a choice between different forms of discipline or different types of conditioning: on the one hand, there is social conditioning, and on the other hand, mental and self-imposed ethical training that transforms a particular "discourse of truth" or "veridiction" (*dire-vrai*) into the very "mode of being of the subject" (Foucault 2005, 327).

There is yet another important alternative that Foucault highlights in relation to ancient ethics: "soul" vs. "life" or, to be more precise, *purification* of the soul vs. *stylistics* of existence. Placing the entire ancient philosophy as well as Christian asceticism under the sign of this fundamental commitment called "care of the self", Foucault, in his final course from 1984, stresses the opposition between the Platonic and Christian understandings of self-care and the (pre-)Socratic and Cynic one. While it remains true that both Platonism and Christianity understand self-care on the grounds of soul–body metaphysical dualism, in the second case, we may observe "this establishment of oneself, no longer as *psukhê* – but as *bios*, no longer as soul but as life and mode of life" (Foucault 2011, 160). You no longer strive for "pure" contemplation; instead, you try to give a permanent "account" of yourself in the light of a fundamental choice for a "mode of existence which is to be examined and tested" throughout your entire life (Foucault 2011, 161). The Cynic alternative does not involve purification of the soul with the prospect of an eternal life; instead, it focuses on the most "natural" style of living accessible to a human being, rejecting the artificiality of social conventions that came to define "normality". And this is how, drawing on the "final" Foucault, we may envisage a possible reshuffling of the meaning of "bioethics" as an *ethics of "life"* (*bios*) conceived as resistance to a *politics* of life or a political seizure of power over one's very existence.

But how are we to make a choice between different ethics or "arts of living"? I state that Foucault's aesthetics of existence seems to imply something different from the priority of decision over norms (as in the embrace of the irrational character of "original choice" in Existentialism), as well as from the idea of steadying definitive traits of character (the goal of traditional virtue ethics). Rejecting the arbitrariness of decisionism brings into play the vital need for small communities of "critical friends" who validate your choices and core commitments (somehow similar to the recognition that an artist gains for his work through the creation of an "audience" for

---

12 In his famous 1945 lecture *L'Existentialisme est un Humanisme*, Sartre had already advanced the claim that we live in an age where "the moral choice is comparable to the construction of a work of art" (Sartre 2001, 35): we cannot rely anymore on rules established once and for all, but our own choices and actions, provided they are coherent, hold an exemplary value that requires the validation of others.

it)[13]. As to the dismissal of everything that ties an individual "to his own identity in a constraining way" (Foucault 1983, 212), I think this might bring forth Nietzsche's idea of "brief habits"[14] as a possible key for reaching some kind of balance between "the demand for stylistic unity of one's existence and the need for self-distancing" (Iftode 2015, 150).

13.5 Final remarks

It is not advisable, on the basis of Foucault's published work and public interventions, to place him in a well-defined position with regard to a field of bioethics that has grown in scope and importance over the past decades. However, it is interesting to know that being a strong advocate of individual freedom, he was very much in favor of personal choice in matters such as assisted suicide (Foucault 2001b, 1075–6), abortion, sexual conduct, and even the use of drugs (Foucault 2015, 112). Judging the "stylization of existence" and the attempt of "self-creation" as the only viable answers to the strategies of biopolitical normalization displayed in our modern societies, Foucault might have very well pleaded for a vegetarianism inspired by ascetical "self-care", as suggested by Tran (2011), and held a general positive view toward the idea of human enhancement. Nothing forbids us to believe that he would have regarded even genetic enhancement as a "technology of the self" fitted for the future. Nevertheless, from a Foucauldian perspective, we have to be aware of this complicate interplay between techniques of domination and techniques of the self: in fact, the origin of the disciplinary technology developed in Western civilization is to be found in the Christian techniques of the self that emerged as monastic practices (Foucault 2014). So we may assume that Foucault would have relentlessly warned us about how easy it is for an autonomous technique of the self to be seized by some institution or converted into a power technique by a State apparatus. The "transhumanist" plea for biomedical *moral* enhancement would have been treated with the utmost suspicion by the French thinker, and his lifelong commitment to social justice would have

13 Unlike the case of interactions with hostile strangers or unconditional admirers, mutual affection and respect between friends may open up a space where you feel safe to give and receive benevolent criticisms and permanently put to test those ideas and attitudes that matter the most in the world for you at a particular time and thus are defining of your identity.

14 "I love brief habits and consider them an inestimable means for getting to know *many* things and states, down to the bottom of their sweetness and bitterness... brief habits, too, have this faith of passion, this faith in eternity... But one day its time is up" (Nietzsche 1974, 236–7).

made him extremely sensitive to the issue of costs and the extensive availability of enhancement technologies.[15]

References

Agamben, Giorgio. 1995. *Homo sacer. Il potere soverano e la vida nuda, I.* Torino: Giulio Einaudi.

Agar, Nicholas. 2004. *Liberal Eugenics: In Defence of Human Enhancement.* Oxford: Blackwell Publishing.

Associated Press. 2016. "Amid war, UN warns Syrians smoking is a grave health risk". Jun. 9. Available at: http://www.businessinsider.com/ap-amid-war-un-warns-syrians-smoking-is-a-grave-health-risk-2016-6?IR=T

Berlin, Isaiah. 1969. "Two Concepts of Liberty." In *Four Essays on Liberty.* Oxford: Oxford University Press.

Bishop, Jeffrey and Jotterand, Fabrice. 2006. "Bioethics as Biopolitics." *Journal of Medicine and Philosophy.* 31 (3): 205-212.

Dardot, Pierre and Laval, Christian. 2009. *La nouvelle raison du monde: Essai sur la société néolibérale.* Paris: La Découverte.

Esposito, Roberto. 2008. *Bíos. Biopolitics and Philosophy.* Trans. Th. Campbell. Minnesota: University of Minnesota Press.

Foucault, Michel. 1964. "La folie, l'absence d'œuvre." *La Table ronde* No. 196: *Situations de la psychiatrie*: 11-21; reprinted as Appendix in Foucault 2006a.

Foucault, Michel. 1978. *The History of Sexuality I: An Introduction.* Trans. R. Hurley. New York: Pantheon Books.

Foucault, Michel. 1983. "The Subject and Power." In H. Dreyfus and P. Rabinow, *Michel Foucault: Beyond Structuralism and Hermeneutics* (2nd ed.). Chicago: Chicago University Press.

Foucault, Michel. 1990. *The Use of Pleasure. The History of Sexuality 2.* Trans. R. Hurley. New York: Vintage Books.

Foucault, Michel. 1995. *Discipline and Punish: The Birth of the Prison.* Trans. A. Sheridan. New York: Vintage Books.

Foucault, Michel. 1997. "The Ethics of the Concern for Self as a Practice of Freedom." In *Ethics, Subjectivity and Truth. Essential Works of Foucault, vol. 1.* Editor Paul Rabinow. New York: New Press: 281-301.

Foucault, Michel. 2001a. "La naissance de la médecine sociale." In *Dits et écrits II: 1976-1988,* Paris: Gallimard: 207-228.

Foucault, Michel. 2001b. "Conversation avec Werner Schroeter." In *Dits et écrits II: 1976-1988,* Paris: Gallimard: 1070-1079.

Foucault, Michel. 2003a. *Abnormal. Lectures at the College de France 1974-1975.* Trans. G. Burchell. London and New York: Verso.

Foucault, Michel. 2003b. *"Society Must Be Defended": Lectures at the Collège de France 1975-1976.* Trans. D. Macey. New York: Picador.

---

15 A part of this article was written as a reply to some objections advanced by Emanuel Socaciu, Radu Uszkai, and Toni Gibea, during the conference on "Bioethics in European Context", University of Bucharest, 19-20 May, 2016. I would also like to thank the editors Emilian Mihailov and Tenzin Wangmo for their careful reading of my draft and all the useful comments and questions they raised.

Foucault, Michel. 2004. *Naissance de la biopolitique. Cours au Collège de France 1978-1979*. Paris: Seuil/Gallimard.

Foucault, Michel. 2005. *The Hermeneutics of the Subject. Lectures at the Collège de France 1981-1982*. Trans. G. Burchell. New York: Palgrave Macmillan.

Foucault, Michel.2006a. *History of Madness*. Trans. J. Murphy and J. Khalfa. London and New York: Routledge/Taylor & Francis e-Library.

Foucault, Michel.2006b. *Psychiatric Power: Lectures at the Collège de France 1973-1974*. Trans. G. Burchell. New York: Palgrave Macmillan.

Foucault, Michel.2008. *The Birth of Biopolitics. Lectures at the Collège de France 1978-1979*. Trans. G. Burchell. New York: Palgrave Macmillan.

Foucault, Michel. 2009. *Security, Territory, Population. Lectures at the Collège de France 1977-78*. Trans. G. Burchell. New York: Palgrave Macmillan.

Foucault, Michel. 2011. *The Courage of the Truth. Lectures at the Collège de France 1978-1979*. Trans. G. Burchell. New York: Palgrave Macmillan.

Foucault, Michel. 2014. *On the Government of the Living. Lectures at the Collège de France 1979-1980*. Trans. G. Burchell. New York: Palgrave Macmillan.

Foucault, Michel. 2015. *The Punitive Society. Lectures at the Collège de France 1972-1973*. Trans. G. Burchell. New York: Palgrave Macmillan.

Foucault, Michel. 2017. *Subjectivity and Truth. Lectures at the Collège de France 1980-1981*. Trans. G. Burchell. London: Palgrave Macmillan.

Gros, Frédéric. 2004. "Michel Foucault, une philosophie de la vérité." In *Michel Foucault: Philosophie (Anthologie)*. Paris: Gallimard.

Iftode, Cristian. 2012. "Philosophy, Terror, and Biopolitics." *Public Reason*. 4 (1-2): 95-105.

Iftode, Cristian. 2015. "The Ethical Meaning of Foucault's Aesthetics of Existence." *Cultura. International Journal of Philosophy of Culture and Axiology* 12 (2): 145-162.

Laing, R.D. 1971. *The Politics of Experience*. New York: Ballantine Books.

Le Blanc, Guillaume. 2006. *La pensée Foucault*. Paris: Ellipses.

Lemke, Thomas. 2011. *Biopolitics: An Advanced Introduction*. Trans. E. Trump. New York: N.Y. University Press.

Liesen, Laurette, and Mary Walsh. 2011. "The Competing Meanings of 'Biopolitics' in Political Science: Biological and Post-Modern Approaches to Politics." In *APSA 2011 Annual Meeting Paper*. Available at: http://papers.ssrn.com/sol3/papers.cfm?abstract_id=1902949

Lombroso, Cesare. 2006. *Criminal Man*. Trans. M. Gibson and N. Hahn Rafter. Durham, NC: Duke Univ. Press.

McGushin, Edward. 2007. *Foucault's Askêsis: An Introduction to the Philosophical Life*. Evanston, IL: Northwestern University Press.

Nietzsche, Friedrich. 1974. *The Gay Science*. Translated by Walter Kaufmann. New York: Vintage Books.

O'Leary, Timothy. 2002. *Foucault and the Art of Ethics*. London and New York: Continuum.

Reiner, Peter. 2013. "The Biopolitics of Cognitive Enhancement." In E. Hildt and A.G. Franke (eds.), *Cognitive Enhancement: An Interdisciplinary Perspective*. Dordrecht: Springer Science.

Revel, Judith. 2002. *Le vocabulaire de Foucault*. Paris: Ellipses.

Sandel, Michael. 2007. *The Case Against Perfection: Ethics in the Age of Genetic Engineering*. Cambridge, MA: Harvard University Press.

Sartre, Jean-Paul. 2001. "Existentialism and Humanism." In *Basic Writings*. Steven Priest (ed.). London and New York: Routledge.

Snowdon, Christopher. 2016. "The UN's warning to Syrian smokers is beyond parody". *The Spectator*, 14 June. Available at: http://blogs.spectator.co.uk/2016/06/uns-warning-syrian-smokers-beyond-parody/

Tran, Jonathan. 2011. *Foucault and Theology*. London: T&T Clark International.

Isabelle Wienand, Milenko Rakic, Sophie Haesen, Bernice Elger

14 How Should One Die? Nietzsche's Contribution to the Issue of Suicide in Medical Ethics

14.1 Introduction

In the history of ideas, "suicide" refers in general to the voluntary act of self-destruction (Minois 1995). By contrast, the word "suicide" is hardly used alone in contemporary medical ethics, as patients' requests to end their lives refers to specific ethical and legal issues (Beauchamp 1993). For instance, "physician-assisted suicide" raises the question about the justified conditions under which physicians may be involved in the request of patients to die. The fundamental condition is that the decision-making of the patient be unambiguously autonomous. A further use of the word "suicide" in medical ethics is within the context of "suicide prevention". It refers to the medical–social right to intervene in order to prevent patients who are mentally ill, or clinically depressed, from not acting autonomously (Beauchamp and Childress 2009). This paper focuses upon the philosophical arguments on assisted suicide, which the German thinker Friedrich Nietzsche (1844–1900) developed in his early writings. We argue that Nietzsche – who is often perceived as defending the legitimacy of suicide as the expression of individual autonomy (Hecht 2013) – remains, however, cautious and undecided concerning the possibility of identifying for certain whether a patient is indeed acting autonomously when s/he requests to put an end to his/her life. We underline the hermeneutic challenges that health-care professionals face when a patient formulates a suicide request. In this paper, we ask whether it should be desirable that health-care professionals suspend their personal view regarding assisted suicide.

The paper focuses on Nietzsche's ideas about suicide, which he formulated at the end of the 1870s, mainly in *Human, all too Human* (HTH). In this contribution, we shall take a closer look at §185 of *The Wanderer and His Shadow* and suggest, in contrast to a widespread view, that Nietzsche is not an active and straightforward defender of the cause of suicide.

Our suspicion is that Nietzsche was aiming at something else than at simply endorsing a pro-suicide attitude. When Nietzsche took up the case of suicide, it was mainly to criticize philosophical rejections of suicide, in particular Schopenhauer's view (Jacquette 2000), the religious prohibition, and the moral condemnation of suicide. The issue of suicide was an opportunity for Nietzsche to distance himself from normative stances. By criticizing the philosophical, moral, and religious condemnation of suicide, Nietzsche did not, however, plead for a legitimation of suicide. The suggestion made here is that Nietzsche's disagreement with the moral condemnation of suicide does not necessarily resolve the issue at stake. Indeed, there is no clear affirmation of suicide in Nietzsche's texts. As we shall see later on, many elements in §185 make it difficult to ascertain Nietzsche's view on suicide.

Unlike Paul Loeb's recent work on Nietzsche's conception of suicide in comparison to Camus's existentialist understanding (Loeb 2010), our purpose here is to focus on earlier texts about suicide, i.e., from the period of HTH. We shall leave out the later texts from the *Zarathustra* and the *Genealogy of Morals*, which Loeb has studied. In difference with Paulo Stellino's recent publication on suicide (Stellino 2013), we are not primarily interested in discussing the possible tension between Nietzsche's apology of suicide and his affirmation of life.

Our interest was raised by contemporary debates on suicide in medical ethics, in which Nietzsche appears now and then as a figure of authority to defend the morally legitimate option of assisted suicide (Benzenhöfer 2009; Filiberti et al. 2001). By and large, the legal and ethical debates oppose those who think that suicide is a fundamental right of the patient and those who defend the view that suicide may not be seen as a right, because it is an irrational act committed out of despair, loneliness, life fatigue, and depression (Mayo 1986).

In contrast, the advocates of medically assisted suicide defend (Küng 2014) the conception of human dignity, according to which dying in dignity means dying before ineluctable pain becomes unendurable, and before the loss of intellectual, emotional, or motoric competences is irretrievable. In these delicate issues, Nietzsche is often cited (mainly the *Zarathustra*) as endorsing the choice of physician-assisted suicide (Pabst Battin 2015). For the opponents of physician-assisted suicide, Nietzsche's legitimation of "reasonable death" is a violation of the principle of sanctity of life (Moreno 1995). The literature often refers to §88 of HTH, in which Nietzsche writes that there is no right to stop someone from committing suicide: "Prevention of Suicide. – There is a right according to which we may deprive a man of life, but none that permits us to deprive him of death: this is merely cruelty" (Nietzsche 1988, vol. 2, 87; our translation).

Of course, it is anachronistic to appeal to Nietzsche in order to defend the particular case of "physician-assisted suicide" within the particular setting of current medical ethics. We are too quick to appropriate him as a timely ally, in order to defend this ethical issue.

Apart from this undue appropriation of Nietzsche's views, there are deeper problems attached to the interpretation, according to which the German philosopher defends actively the cause of suicide in §185. We shall mention and briefly develop two problems: (1) the first and most interesting one is that Nietzsche does not clearly indicate his view on suicide in §185. The hermeneutic challenge of this text makes it impossible to work out a clear position for or against suicide. This is not necessarily to be interpreted as a conceptual weakness of Nietzsche's arguments. On the contrary, we tend to think that Nietzsche's undecidedness speaks for his nondogmatism and moral sensitivity. (2) The second is that Nietzsche is mainly aiming at criticizing dominant moral and religious condemnations of suicide. Indeed, his main argument is that these opinions reject individual autonomy. This second problem is only briefly mentioned here. We shall conclude by a biographical testimony written by Nietzsche's colleague and friend, Franz Overbeck. The Basler Professor for History of Christendom, Overbeck remembers in his *Memoirs on Friedrich Nietzsche* (Overbeck 1906) that Nietzsche was very preoccupied with the ideal of suicide. Overbeck even quotes HTH §185 as an illustration for the philosophical and personal importance of this issue in his friend's life. Let us start with the first problem, the hermeneutic challenge of §185.

14.2 The hermeneutic challenge of HTH §185

HTH §185 reads as follows:

> Of Reasonable Death. – Which is more reasonable, to stop the machine when it has done the job demanded of it, or to let it run until it stops on its own – in other words, until it is deteriorated? Is not the latter a waste of the maintenance costs, a misuse of the strength and care of the operators? Are we not here throwing away something which would be necessary elsewhere? Are we not propagating a kind of contempt of the machines, in the sense that many of them are so uselessly maintained and operated? – I am speaking of involuntary (natural) and voluntary (reasonable) death. Natural death is independent of all reason and is really an *unreasonable* death, in which the pitiable substance of the shell determines how long the kernel should exist or not; in which, accordingly, the deteriorating, ailing and dull jailer is lord and indicates the moment at which his noble prisoner shall die. Natural death is the suicide of nature – in other words, the annihilation of the rational being through the irrational being that is attached thereto. Only a religious perspective can make the reverse appear; for then, as is equitable, the higher reason (God) issues its orders, which the lower reason has to obey. Outside religious thought pattern natural death is not worth glorifying. – The wise dispensation and disposal of death belongs to that now quite incomprehensible and immoral-sounding morality of the future, whose dawn must be an indescribable bliss to behold. (Nietzsche 1988, vol. 2, 632–3)

One tends to read Nietzsche's early texts prospectively. Sometimes, it is useful. Often it biases one's understanding. A good illustration for our odd reading is the issue of suicide. Many commentators read §185 with the chapter of the *Zarathustra* "Of Voluntary Death" (Nietzsche 1969, 97–9) in their mind. For instance, Andreas Urs Sommer starts his commentary of §185 by writing that in many regards, *The Wanderer and His Shadow* anticipates the motto of the *Zarathustra*, "die at the right time" (Sommer 2010, 171). Sommer is certainly right in establishing some connections between, on the one hand, the idea of the free spirits, namely, shaping one's life according to one's own rules, and the wise Zarathustra on the other. After all, §185 might be also a plea for considering death as it is, without fear, without the metaphysics of tragedy, without religious consolation, and without moral condemnation. So, if death is part of life, i.e., if death constitutes the normal, banal, trivial, *natural* fate of human existence, then one should consider it rationally and take reasonable decisions about one's life, when it seems to be fading away. But the arguments get more complicated in §185. Could Nietzsche be arguing 1) that death is a natural event, *and* – as Nietzsche does in §185 – 2) that we should *not* die in a natural way? What does "natural" mean in 1) and 2)? We suspect the question we should ask concerning §185 is not what position is Nietzsche endorsing, but what suicide actually means.

To start with, the word "suicide" appears only once in §185, and in a very particular phrase: "Suicide of nature" (*Selbstmord der Natur*). In the common use of the word (putting an end to one's own life), it is noteworthy that "suicide" is neither in the title, nor in the aphorism, nor in the preliminary version of the published text (*Vorstufe*): "There is no right by which we can prevent a human being from taking his/her life. To put the criminal in this position of 'ought-to-live' is cruelty" (Nietzsche 1969, vol. 4/4, 182; our translation). Nietzsche could have well used it here, since he uses the word "suicide" (*Selbstmord*) in HTH I §88.

Second, Nietzsche begins §185 by asking a series of questions, four long questions in total. The questions do not actually deal with the issue of suicide *sensu stricto*. Instead, Nietzsche is asking whether it is more reasonable to repair an already old machine and to run it until it stops, or to stop it (*stillstellen*) before it starts breaking down. One could argue here that Nietzsche speaks about suicide *per analogiam*. Fair enough, Nietzsche discusses the question of stopping at the right time, on due course, out of respect, and so on. Nietzsche also explicitly confirms his analogy by writing: "I am speaking of involuntary (natural) and voluntary (reasonable) death". To our understanding, the analogy does not really clarify the issue of death/suicide, and we think that Nietzsche was aware of the obscurity of the analogy too. He would not have otherwise reminded his readers of the topic in question: "I am speaking" (*ich spreche*). The analogy is indeed very puzzling. What "machine" should one stop when it is time to do so: one's body, one's soul or oneself, or something else? Nietzsche uses the word "machine" in different contexts, sometimes to explain something about humans. Consider, for instance, this posthumous text from 1876–1877: "Mankind, a disorderly functioning machine with formidable strength" (*Menschheit, eine unordentlich fungierende Maschine mit ungeheuren Kräften*) (Nietzsche 1988, vol. 8, 369; our translation). In other texts, similar to that depicted here, "machine" seems to refer to the mechanistic Cartesian concept of body (1985, 99-108). He could also allude to La Mettrie's materialism (La Mettrie 1996).

The unsettling character of the analogy lies also in the unexpected language Nietzsche uses: "waste" (*Vergeudung*), "maintenance costs" (*Unterhaltungskosten*), "operators" (*Bedienende*), and "throwing away" (*wegwerfen*). All these words and kinds of arguments are typical for the modern, industrialized, capitalist 19th-century Europe (Thomas 1983). The negative questions that Nietzsche is raising are directly connected with the impact of the economy of production and consumption on culture, morality, and religion. Consider, for instance, this question: "Are we not propagating a kind of contempt of the machines, in the sense that many of them are so uselessly maintained and operated?"

Even if we agreed that the machine represents the human being as a whole, it would still remain unclear to know for sure *what position is more reasonable* to Nietzsche's eyes. Indeed, Nietzsche suggests two options but formulates objections *only to one option*: option a), namely, stopping the machine before it deteriorates would mean throwing away resources that could still be of use. As for option b), namely, maintaining the machine at all costs, Nietzsche does not express his opposition. It does not mean, however, that Nietzsche supports this option. In sum, the analogy between humans and machines has not clarified the philosophical position of Nietzsche. Nietzsche seems to be endorsing an external observer looking at the industrial revolution in the late 19th century Europe and considering its impact on our moral and religious lives. But we still have learned very little about Nietzsche's own position on suicide!

Third, Nietzsche does not clarify his position when he is supposed to. The reader who expects an explanation of the analogy will be disappointed. For Nietzsche does not explain the meaning of "machine". What is more, Nietzsche introduces a heavy-weighted philosophical pair of opposed concepts, namely, "involuntary/voluntary", without explaining them. The reader is at sea with the juxtaposition of "involuntary" with "natural". The oddest pair is certainly the opposition "reasonable/natural". Does Nietzsche refer to, and possibly endorse, the Stoic position on suicide, as Stellino indicates (Stellino 2013)? Stellino is right to underline the important influence of the Ancient Stoic position on suicide, with which Nietzsche was indeed very familiar. However, there is no explicit element in §185 which indicates that Nietzsche acquiesces to it. Furthermore, there are texts, for instance, the *Enchiridion* by Epictetus, wherein the Roman philosopher argues that it is "unreasonable to leave", as it does not follow the rational principle to live according to nature (*secundum naturam vivere*) (Epictetus 2004, 16-17).

The obscurity of §185 reaches its climax when Nietzsche explicates why natural death is unreasonable. Here too, the language he uses can easily mislead the reader. Indeed, Nietzsche uses the metaphysical language of dualism (shell/kernel) to show that the shell, the body ("the pitiable substance") determines the life of the soul, the kernel. Nietzsche is using the model of platonic dualism *ad absurdum* in order to demonstrate the primacy of the body over the soul. By doing so, Nietzsche seems to argue that the only kind of death *at our disposal is natural death*. Death cannot possibly be decided by reason. Nature decides when the soul dies. But if it is true that there is only natural death, and no reasonable death, why speak of "reasonable death" at all if we may not choose? One possible way out of this apparent contradiction is to distinguish between *reasonable death* and *dying reasonably*. For Nietzsche, dying reasonably would be to accept the power of nature. Furthermore, as Sommer suggests, choosing the option of "dying in a reasonable way" implies giving up the belief in the immortality of the soul and the body (Sommer 2010). Dying reasonably or philosophically would imply accepting one's finitude. However, Nietzsche doubts whether humans can give up their belief in a postmortem life so easily, as the matter of dying is not only a matter of reason. This is where the second problem appears: our religious, moral, and metaphysical expectations. We shall concentrate our attention on the religious view, particularly the Christian conception of death, as it is the main focus of Nietzsche's critical analysis in §185.

14.3 The religious interpretation of suicide

In §185, religion seems to be the root of the problem. More precisely, the religious way of looking at our life, as well as at our death, creates an obstruction, so Nietzsche argues. And here, "religious" should not be understood in a strict confessional, or theological way, although, by "religious", Nietzsche often means a Christian-based view. "Religious" refers to a typical moral view consisting in claiming that the natural world as it is – i.e., creature of lower ontological and moral value – needs to be changed, in order to be saved by God. Let us transpose this dualistic construction onto the issue of suicide: religion claims that natural death is not something trivial that we ought to accept. Furthermore, it is God qua *higher reason* who decides upon the humans' lives qua *lower reason*. The religious perspective inverts the relation between nature and reason: divine reason is not submitted to the laws of nature. As the creator of all, God gives and takes away human life. Humans are not considered autonomous beings (e.g., 1 Samuel 2, 6; John 1, 3–4).

There is one further point to elucidate. If religion conquers nature, why is it only within the religious perspective that "natural death" is glorified? "Natural death" does not appear as a trivial phenomenon, for it is decided by divine commandment. One would commit a sin against nature, if one were to decide one's own time of death. Unlike this religious naturalism, which sees in "natural death" the sign of God's almighty power, Nietzsche sees suicide qua "natural death" as a human phenomenon.

To sum up, the religious perspective interprets natural death as the expression of divine reason. Nietzsche's skeptical objection is to say that we do not have sufficient knowledge about death to claim essential links between death and divine reason. This might help us find out Nietzsche's position. As hinted at earlier, we suggest that Nietzsche could opt for "dying philosophically" or "wisely": we should recognize the power of nature over us. In our view, Nietzsche cannot be assigned a clear pro or anti position on suicide. His challenging way of posing the problem of suicide helps us articulate the conception of nature and reason in today's debates in medical ethics – e.g., also Nietzsche's polemical aphorism §36 "Moral for doctors" in his late work *Twilight of the Idols*:

> Here it imports above all – in spite of all the cowardice of prejudice – to establish, the right, that is, the physiological, appreciation of the so-called *natural* death, which is ultimately also "unna-

> tural" death, suicide. One never perishes through anyone else but oneself. But this [natural] death happens under the most miserable conditions, an unfree death, death not at the right time, a coward's death. *Out of* love for life one should want death differently: free, conscious, without fortuity, without invasion. (Nietzsche 1988, vol. 6, 135; our translation)

The seemingly inevitable confusion over the issues of suicide is well illustrated in the above-quoted passage: "natural" death ends up being "unnatural". The casuistic vocabulary related to death in modern medical ethics equally indicates similar difficulties in determining the ethical and legal framework of physician-assisted suicide (Foot 2002).

If we limit the scope to a medical context, complex ethical issues, such as the concept of patients' autonomy and accountability, ethics of care, free will, or the possibility and the limits of overtreatment need to be considered. The essential issue of the goals of medicine should not be forgotten either (WHO 2002). Opponents of the notion that suicide can be justified often argue that suicide only happens when an individual is depressed (Hecht 2013), thus dismissing the possibility that suicide can be the expression of individual autonomy (Battin 2015). This seems to be an ongoing conflict between palliative care professionals claiming that sufficient palliative care eliminates any reason for a patient wanting to end his or her life on the one hand and advocates of the right-to-die movement for whom patient autonomy is to be respected on the other hand (Nitschke and Stewart 2013).

Can something be done to find an alternative to this unfruitful conflict? Nietzsche's nondogmatic position toward suicide might be helpful, inasmuch as he reminds that patients' autonomy should be linked with the recognition of patients' vulnerability – or as Nietzsche writes, with the awareness of the power of nature over us. Furthermore, Nietzsche's apparent ambiguity on the issue of suicide can be read as an attempt to make room for a number of defendable positions pro and contra suicide. He seems to be suggesting that suicide is a vexed question in modern society, because we continue to argue within the theological, religious frame, without even noticing it: we claim autonomy and freedom of thought but remain in many subtle ways submitted to moral and religious imperatives. Beyond Nietzsche's unflattering portrayal of modernity, it seems that he insists upon the task – how difficult it may be – to attend to the individual's wish to die. And this seems to us an important point for the medical team's attitude to patients' wish to die: assisted suicide should not be about the personal beliefs and morals convictions of the medical team or the patients' family, but about respecting patients' autonomy. The hermeneutic tasks that health-care professionals face when a patient formulates a suicide request are often challenging, as they might be opposed to their own moral principles and/or religious beliefs. A paradigmatic example in the literature is the discussion of the conditions under which the Hippocratic Oath allows assisted suicide (Veatch 2012; Weithman 1999). Therefore, it might be desirable for them to suspend their dogmatic position regarding assisted suicide when dealing with autonomous patients' requests. It might even help to recall Nietzsche's suggestion quoted above, that "out of love for life" patients might want "death differently".

14.4 Conclusion

This paper contends that Nietzsche is unambiguously endorsing suicide in this often quoted text of §185 *The Wanderer and His Shadow*. We identify two major problems in seeing Nietzsche as an explicit proponent of suicide. First, the undecided, conditional position of Nietzsche in §185 makes it difficult to consider him a militant defending the right to suicide. We argued that the cautious, if not skeptical, position of Nietzsche should not be read as a conceptual insufficiency, but as a salutary nondogmatic position. Second, Nietzsche's primary concern is not to endorse or reject the legitimacy of suicide per se. His focus is much more on calling into question the religious interpretation of suicide, as it does not acknowledge individual autonomy. Finally, the testimony of Overbeck illustrates that Nietzsche's foremost concern is to defend the philosophical legitimacy of autonomy. The fact that he died naturally does not preclude his argument. After all, "reasonable" and "natural" death can both be interpreted as expressions of individual autonomy:

Nietzsche idealized suicide as the 'reasonable death' and gave it the highest recognition in the morality of the future (*The Wanderer and His Shadow*, §185). And under the impression of such statements and similar ones which I heard from him at more than one occasion and which came quasi naturally out of his engagement with the antique world, I thought quite often of suicide as the end granted to him, even with growing conviction, at least until the winter 1883, during which Wagner died and Nietzsche's letters caused me extreme worries in this regard. I almost never thought of madness, or at least very late, shortly before the catastrophe. (Overbeck 1906, 214)

Nietzsche's position toward the ethical issue of suicide contributes in many ways to the rich European tradition starting with the Ancient Stoics. Nietzsche's challenging views also show that the issue of suicide is complex and part of a larger *Weltanschauung*.

References

Battin, Margaret Pabst ed. 2015. *The Ethics of Suicide. Historical Sources*, New York: Oxford University Press.

Beauchamp, Tom. 1993. "Suicide." In *Matters of Life and Death*, edited by T. Reagan. New York: Random House, 69-120.

Beauchamp, Tom, and James Childress. 2009. *Principles of biomedical ethics*. Oxford University Press.

Benzenhöfer, Udo. 2009. *Der gute Tod? Geschichte der Euthanasie und Sterbehilfe*, Göttingen: Vandenhoeck & Ruprecht.

Descartes, René. 1985. *The Philosophical Writings of Descartes*, 3 vols., vol.1, translated by J. Cottingham, R. Stoothoff and D. Murdoch, Cambridge: Cambridge University Press.

Epictetus. 2004. *Enchiridion*, edited by T. Crawford, Mineola NY: Dover Thrift Editions.

Filiberti, Antonio, Carla Ripamonti, Amadio Totis, Vittorio Ventafridda, Franco De Conno, Paolo Contiero, Marcello Tamburini. 2001. "Characteristics of terminal cancer patients who committed suicide during a home palliative care program." *Journal of Pain & Symptom Management* 22: 544-553.

Foot, Philippa. 2002. *Virtues and Vices*, Oxford: Clarendon Press.

Hecht, Jennifer Michael. 2015. *Stay: a History of Suicide and the Philosophies Against It*, New Haven: Yale University Press.

Jacquette, Dale. 2000. "Schopenhauer on the ethics of suicide." *Continental Philosophy Review* 33: 44-58.

Küng, Hans. 2014. *Glücklich sterben?*, Munich: Piper Verlag.

La Mettrie, Julien Offray de. 1996. *Machine Man and Other Writings*, edited by A. Thomson, Cambridge: Cambridge University Press.

Loeb Paul S. 2010. Suicide, Meaning and Redemption, in *Nietzsche on Time and History*, edited by M. Dries, Berlin: de Gruyter, 163-190.

Mayo, David J. 1986. "The Concept of Rational Suicide" *Journal of Medicine & Philosophy* 11(2): 143-155.

Minois, Georges. 1995. *Histoire du suicide. La société occidentale face à la mort volontaire*, Paris: Fayard.

Moreno, Jonathan D. ed. 1995. *Arguing Euthanasia*. The Controversy over Mercy Killing, Assisted Suicide and the "Right to Die". New York: Simon & Schuster.

Nietzsche, Friedrich. 1988. *Sämtliche Werke, Kritische Studienausgabe in 15 Bänden*, G.Colli, M. Montinari (Eds), Berlin/München: DTV/de Gruyter.

Nietzsche, Friedrich. 1969. *Kritische Gesamtausgabe. Nachbericht Zur Vierten Abteilung*. Berlin: de Gruyter, here vol. 4/4, 182 [10], 182.

Nietzsche, Friedrich. 1969. *Thus Spoke Zarathustra: a book for everyone and no one*, translated and introduced by R.J. Hollingdale. Harmondsworth: Penguin.

Nitschke, Philip, H., F. Stewart, F. 2013.*The Peaceful Pill Handbook*. Bellingham: Exit International US.

Overbeck, Franz. 1906. "Erinnerungen an Friedrich Nietzsche". *Die Neue Rundschau*. XVIIter Jahrgang der Freien Bühne, Bd.1. Berlin: Fischer Verlag, 209-231.

Sommer, Andreas Urs. 2010. "Das Sterbendenken. Zur Möglichkeit einer *ars moriendi* nach Nietzsche und Elias" in *Zur Genealogie des Zivilsationsprozesses. Friedrich Nietzsche und Norbert Elias*, edited by F. Günther et al. Berlin: De Gruyter, 159-174.

Stellino, Paolo. 2013. "Nietzsche on Suicide". *Nietzsche-Studien* 42: 151-177.

Thomas, Richard Hinton. 1983. *Nietzsche in German Politics and Society, 1890-1918*. Manchester: Manchester University Press.

Veatch, Robert M. 2012. *Hippocratic, Religious, and Secular Medical Ethics. The Points of Conflict*. Washington, DC: Georgetown University Press.

Weithman, Paul J. 1999. "Of Assisted Suicide and "The Philosophers' Brief" *Ethics*, Vol. 109, No. 3: 548-578.

World Health Organization. 2002. *National cancer control programmes: policies and managerial guidelines 2nd ed.* Geneva: World Health Organization.

Contributors

Octavian Buda is chair, Professor of history of medicine at "Carol Davila" University of Medicine and Pharmacy in Bucharest, Romania. He is Ph.D. postgraduate (medical sciences), since 2001 and Master of Arts in Philosophy at the University of Bucharest, 1997. Between 2015-2017, elected President of the European Association for the History of Medicine and Health (EAHMH), Strasbourg. General research areas: history of medicine / history of epidemics; history of criminology / forensic psychoanalysis; forensic psychiatry / social history of forensic medicine and psychiatry; cultural and psychiatric anthropology / medical epistemology; bioethics / intellectual history. Published books (selection): *On Regeneration and Degeneration of a Nation: Medical Discourses in Bucharest, 1872-1912*, single editor (Bucharest, 2009), 2nd Edition: 2013, Bucharest; *National Identity and Social Medicine: Cultural Anthropology, Psychiatry, and Forensic Medicine in Romania: 1800-1945* (Bucharest, Romanian Academy, 2013). Selected chapters: Medical Enlightenment and Variolation from the Balkans to Romanian territories - to Western Europe, 1672-1802, in: C. Barbulescu and A. Ciupala (Eds): *Medicine Hygiene and Society from the 18th to the 20th centuries*, Ed. Mega, Cluj Napoca, 2011; Historical Aspects of Legal Medicine in Romania in: B. Madea (ed.): *History of Forensic Medicine*, Lehmanns Media, Berlin 2017.

Mihaela Constantinescu is a researcher at the Research Centre in Applied Ethics, Faculty of Philosophy, University of Bucharest, Romania. She obtained her PhD in 2013 at the University of Bucharest, with a thesis on "The Moral Responsibility of Organisations", partly documented at the Rotterdam School of Management, Erasmus University. Her research interests include organizational ethics, virtue ethics, and corporate responsibility. Mihaela co-authored the Romanian volume *Institutionalising Ethics: Mechanisms and Instruments* and is author and co-author of articles published in international volumes and scientific journals such as the "Journal of Business Ethics". With an academic background in both philosophy and communication, Mihaela has worked as a public relations consultant in the private, nongovernmental and public sector, while offering ethics counselling to the latter.

Eva De Clercq has been working as a postdoctoral researcher at the Institute for Biomedical Ethics of the University of Basel (Switzerland) since September 2013. She obtained her PhD in Philosophy from the Catholic University of Leuven (Belgium) and of the University of Pisa (Italy) in 2010. She has a master in social and cultural anthropology of the Catholic University of Leuven (Belgium) and a master in cultural communication of the University of Florence (Italy). Her research interests and expertise include: pediatric palliative care, capacity and decision-making; vulnerability, embodiment and lived-experience and gender related issues. She has published more than 20 peer-reviewed manuscripts in the field of philosophy, bioethics and gender studies. She has published her PhD thesis for the Academic

Publisher Palgrave Macmillan in 2013. She is a lecturer in philosophy and department supervisor at the University Istituto Internazionale Lorenzo de' Medici, Florence (Italy) since 2011.

Bernice Elger is head of the Institute for Biomedical Ethics at the University of Basel and associate professor at the Center for Legal Medicine at the University of Geneva where she is head of the Unit for Health Law and Humanitarian Medicine. She studied medicine and theology in Germany, France, Switzerland and the US and obtained her specialty degree in internal medicine (FMH) in Switzerland. As part of her work at the Center for Legal Medicine at the University of Geneva, she has worked in prison medicine for 15 years. She has been member of the advisory committee of the Chernobyl Biobank for many years and wrote her habilitation thesis on ethical and legal issues of biobanks.

Victoria Federiuc is a doctoral researcher of the State University of Medicine and Pharmacy „Nicolae Testemiţanu", Republic of Moldova. She holds a Pharm. D and MA in Medical ethics and Bioethics from the same university. Her main research topics are pharmaceutical ethics and the ethics of the innovative medical technologies. Victoria is also a teaching assistant at the Department of Philosophy and Bioethics. Other research interests include palliative care, philosophy of medicine and behavioural sciences.

Mihaela Frunză is Associate Professor at the Department of Philosophy, Faculty of History and Philosophy, Babeş-Bolyai University, Cluj-Napoca, Romania. Her research interests include biomedical ethics, philosophy of communication and philosophy for children. She teaches courses in ethics, applied ethics, and philosophy of communication. She has published four books and more than 40 articles in specialized journals.

Sandu Frunză has a PhD in Philosophy of Culture, Values and History. He is PhD coordinator in the field of political sciences. His fields of interests are: deontology, relational ethics, philosophy of religions, Post-Holocaust philosophy, religious fundamentalism, political communication, personal development and ethical counselling. He is editor of the Journal for the Study of Religions and Ideologies. He has recently authored the following books: *Advertising constructs reality. Religion and advertising in the consumer society* (Tritonic, 2014), *Publicitatea și administrația publică sub presiunea eticii* (Lumen, 2015), *Între moartea politicii și moartea lui Dumnezeu* (Eikon, 2017). He also performs philosophical (existential) counselling. His blog address is https://frunzasandu.wordpress.com/

Iulia Grad is teaching assistant at the Department of Communication, PR and Advertising, Faculty of Political, Administrative and Communication Sciences,

Babeş-Bolyai University, Cluj-Napoca, Romania. She obtained her PhD in philosophy in 2012, with a thesis on Martin Buber. Her interest areas are communication ethics, communication philosophy and media communication. The courses and seminars she teaches are: ethics in PR and advertising, advertising, religion, ideology and strategies of dialogue.

Ovidiu Grad is teaching assistant at Iuliu Haţieganu University of Medicine and Pharmacy, Cluj-Napoca, where he teaches surgical semiology. He is also a vascular and general surgeon. Since 2013, Ovidiu Grad is medical doctor, the title of his PhD thesis being "Therapeutic angiogenesis with stem cells in critical lower limb ischemia".

Christopher Gyngell is a Marie Sklodowska-Curie Fellow with the Uehiro Centre for Practical Ethics. His research interests lie primarily in bioethics, moral theory, and the philosophy of health and disease. He is currently working on a Marie Sklodowska-Curie Fellowship funded project titled "Selecting, Creating and Modifying Embryos", which will investigate the ethical and legal implications of new reproductive technologies, such as the gene editing technique CRISPR.

Sophie Haesen has a background in Medicine and Epidemiology (M. Sc. from the Swiss Tropical and Public Health Institute/University of Basel). Among others, she participated at "Ecohealth 2014" in Montreal, "Paratrop 2014" in Zurich, the "63rd Annual Meeting of the American Society of Tropical Medicine and Hygiene" in New Orleans and "Environment and Health" in Basel presenting research projects on synergies resulting from joint human, animal and plant health services and on estimating the disease burden of noma (cancrum oris). Sophie is currently a PhD candidate at the Institute of Biomedical Ethics, University of Basel, Switzerland. Her research project, which is funded by the Swiss National Foundation, focuses on the mental health of older prisoners in Switzerland. Other research interests include general prison health as well as questions of autonomy in Public Health and around the end of life.

Marcello Ienca is research fellow at the Health Ethics & Policy Lab, Department of Health Sciences and Technology at ETH Zurich, Switzerland. His research focuses on the ethics and governance of biomedical data, ethically aligned artificial intelligence (AI) and responsible innovation for emerging technologies at the human-machine interface. Prior to joining ETH, Dr. Ienca has obtained MSc, MA and PhD degrees from the Humboldt University of Berlin, KU Leuven (Erasmus Mundus Consortium) and the University of Basel. He has published extensively on the ELSI of neurotechnology and AI, big data trends in biomedicine, ethical design in robotics, data ethics, dual-use, digital health and cognitive assistance for people with intellectual disabilities. He has advocated the development of neuro-specific rights for the era of neurotechnology and big data. Ienca has received several awards for social responsibility in science and technology such as the Prize Pato de Carvalho (Portugal) and the Paul Schotsmans Prize from the European Association of Centres of Medical Ethics (EACME). He has authored +20 scientific articles in peer-review journals, several book chapters and is a frequent contributor to Scientific American. Ienca is a representative to the board of the International Neuroethics Society and the coordinator of the Swiss Network of Neuroscience, Ethics and Law (SNNEL).

Cristian Iftode (born in 1979) is Associate Professor at the Faculty of Philosophy, University of Bucharest. His published books and edited volumes in Romanian include: *Critique, Marginality, Cynicism* (co-edited with C. Voinea, University of Bucharest Publishing House, 2017); *Aristotle: The Problem of Analogy and the Philosophy of Donation* (University of Bucharest Publishing House, 2015); *Philosophy as a Way of Life: Sources of Authenticity* (Paralela 45, 2010). He is also the author of an important number of academic studies published in Romanian, English and French, dealing with topics in the history of philosophical ideas, ethics and social philosophy, aesthetics, and French post-structuralism. Contact: cristian.iftode@filosofie.unibuc.ro.

Blanca Rodríguez López is Associate professor of moral and political philosophy at the Complutense University of Madrid. At the beginning of her career she worked on utilitarianism, rational choice theory and game theory. Later she worked on liberalism and social norms and in the last few years she has focused on bioethics and human enhancement. Her work is currently focused on moral enhancement, and its relation with cognitive and mood enhancement. She also works on experimental ethics.

Emilian Mihailov is the Executive Director of the Research Centre in Applied Ethics (CCEA) and Lecturer at the Faculty of Philosophy, University of Bucharest. He was a postdoctoral fellow of the Romanian Academy, Iasi Branch, and visiting researcher at Oxford Uehiro Centre for Practical Ethics, University of Oxford. Among Emilian's research interests are neuroethics, moral psychology, Kantian ethics, analytic moral philosophy, applied ethics, and the evolution of morality. Currently he is working on the implications of experimental moral psychology and neuroscience for normative and applied ethics. His book, *The Architectonics of Morality* (Paralela 45 Publishing House 2017), develops a criticist reconstruction of Kantian ethics.

Bogdan Olaru has studied philosophy in Iași (Romania) and Berlin, and graduated psychology at University of Göttingen (Germany). His primary areas of interest include moral philosophy and moral psychology, social justice, bioethics, and phenomenology. His most recent publications address issues related to responsibility and social justice. He edited the anthologies *Autonomy, Responsibility, and Health Care: Critical Reflections* (Bucharest, 2008) and *Contributions to Moral Psychology: Evaluating Results and New Empirical Research* (Bucharest, 2015, with Andrei

Holman), and published *Changing one's Moral Luck: The Effect of Restitution on Moral Judgments* (AV Akademikerverlag, 2012).

Adriana Paladi is an Associated professor at the Philosophy and Bioethics Department of the State University of Medicine and Pharmacy "N. Testemitanu" (Republic of Moldova), and the expert member of National Ethics Committee for clinical research of drugs and new methods of treatment. Her current work lies at the intersection of ethical theory and practice, with a particular focus the value of the informed consent in biomedical research, the exploration of the moral conditions for carrying out a biomedical research without an informed consent, the determination of the ethical problems of medical research on vulnerable populations, ethical issues in end of life care, suffering in end of life care, psychological and spiritual assistance of dying etc.

Milenko Rakic is PhD candidate at the Institute of Biomedical Ethics, University of Basel, Switzerland, with a background in Medicine and Biomedical Sciences (M. Sc. from the University of Bern, 2014). Milenko's research project focuses on self-determination of palliative care patients and is funded by the Swiss Academy of Medical Sciences. His research interests include patients' self-determination in palliative care as well as in clinical trials with cutting edge biotechnology. He was trained in both qualitative and quantitative research. Additionally, he is doing a Master in Health Promotion at the Swiss Distance University of Applied Sciences, Switzerland.

Daniela Rusnac is teaching assistant at the Department of Philosophy and Bioethics at State University of Medicine and Pharmacy „Nicolae Testemiţanu", Chisinau, Republic of Moldova. Daniela is a graduate in philosophy and holds a MA degree in Philosophy. Her research area includes ethical aspects of biotechnology, human enhancement, ethical and humanitarian expertise.

Laurenţiu Staicu is Professor at the Faculty of Philosophy, University of Bucharest, where he teaches Metaphysics, Philosophy of Biology and Ethics of Science. His most recent book, *Three Metaphysical Histories for Insomniacs* (Trei Publishing House, Bucharest, 2017), addresses a number of classical metaphysical puzzles from the Western tradition - personal identity, the reality of the external world, the nature of existence – and tries to offer a fresh perspective over these issues by making a strong connection between traditional philosophical perspectives and their contemporary scientific counterparts.

Constantin Vică is Assistant Professor at the Faculty of Philosophy and researcher at the Research Centre in Applied Ethics, University of Bucharest. His main academic interest are computer and information ethics, roboethics, philosophy of computer science, social and political philosophy, and the critique of intellectual property.

He published several articles and studies on information and ownership, intelligent assistive technologies, online trust, web search engines ethics, digital dialectics, pirate politics, evolution of programming languages, and free software, authorship and intellectual property. Some recent publications: "Kynics and cynics within the political game of information" (with Cristina Voinea, in Critică, marginalitate, cinism, 2016), "Space and Time in the Networked Digital World. A Kantian Critical Examination" (Revista de filosofie, 2016), and "Life and Ownership: Intellectual Property limits, from Genes to Synthetic Biology" (in Ethics in Biology, Engineering and Medicine, 2015). He also co-edited one book: *Filosofia științelor umane. In memoriam Mihail Radu Solcan* (2015, University of Bucharest Press). Contact: constantin.vica@filosofie.unibuc.ro.

Tenzin Wangmo is a Senior Researcher at the Institute for Biomedical Ethics at the University of Basel. She completed her doctoral education in Gerontology (United States) and habilitation in medical ethics (Switzerland). Her research works for the last ten years have encompassed topics such as older prisoners, end-of-life, elder abuse, and intergenerational relationships. Currently, her bioethics research includes technology and aging, empirical bioethics, healthcare data sharing, and older prisoners' rights and access to healthcare.

Isabelle Wienand is a postdoctoral researcher at the Institute for Biomedical Ethics at the University of Basel. She studied philosophy and German literature at the Sorbonne University Paris-IV. Her area of research is modern moral philosophy – particularly Descartes and Nietzsche –and more recently, ethics and philosophy of medicine.